CHEMISTRY RESEARCH AND APPLICATIONS

AN INTRODUCTION TO ELECTRONIC STRUCTURE THEORY

CHEMISTRY RESEARCH AND APPLICATIONS

Additional books and e-books in this series can be found on Nova's website under the Series tab.

CHEMISTRY RESEARCH AND APPLICATIONS

AN INTRODUCTION TO ELECTRONIC STRUCTURE THEORY

NADIA T. PAULSEN
EDITOR

Copyright © 2020 by Nova Science Publishers, Inc.

All rights reserved. No part of this book may be reproduced, stored in a retrieval system or transmitted in any form or by any means: electronic, electrostatic, magnetic, tape, mechanical photocopying, recording or otherwise without the written permission of the Publisher.

We have partnered with Copyright Clearance Center to make it easy for you to obtain permissions to reuse content from this publication. Simply navigate to this publication's page on Nova's website and locate the "Get Permission" button below the title description. This button is linked directly to the title's permission page on copyright.com. Alternatively, you can visit copyright.com and search by title, ISBN, or ISSN.

For further questions about using the service on copyright.com, please contact:
Copyright Clearance Center
Phone: +1-(978) 750-8400   Fax: +1-(978) 750-4470   E-mail: info@copyright.com.

NOTICE TO THE READER

The Publisher has taken reasonable care in the preparation of this book, but makes no expressed or implied warranty of any kind and assumes no responsibility for any errors or omissions. No liability is assumed for incidental or consequential damages in connection with or arising out of information contained in this book. The Publisher shall not be liable for any special, consequential, or exemplary damages resulting, in whole or in part, from the readers' use of, or reliance upon, this material. Any parts of this book based on government reports are so indicated and copyright is claimed for those parts to the extent applicable to compilations of such works.

Independent verification should be sought for any data, advice or recommendations contained in this book. In addition, no responsibility is assumed by the Publisher for any injury and/or damage to persons or property arising from any methods, products, instructions, ideas or otherwise contained in this publication.

This publication is designed to provide accurate and authoritative information with regard to the subject matter covered herein. It is sold with the clear understanding that the Publisher is not engaged in rendering legal or any other professional services. If legal or any other expert assistance is required, the services of a competent person should be sought. FROM A DECLARATION OF PARTICIPANTS JOINTLY ADOPTED BY A COMMITTEE OF THE AMERICAN BAR ASSOCIATION AND A COMMITTEE OF PUBLISHERS.

Additional color graphics may be available in the e-book version of this book.

Library of Congress Cataloging-in-Publication Data

ISBN: 978-1-53618-411-2

Published by Nova Science Publishers, Inc. † New York

CONTENTS

Preface		vii
Chapter 1	Resultant Information Approach to Donor-Acceptor Systems *Roman F. Nalewajski*	1
Chapter 2	Electronic Structure of Solid Oxides Doped with Transition Elements *N. Chezhina and D. Korolev*	59
Chapter 3	Mathematical Modeling of Electronic Structure of Some Nanomaterials *Azad A. Bayramov and Arzuman G. Gasanov*	99
Chapter 4	Basics and Applications of Electronic Structure Theory *Aditya M. Vora*	131
Chapter 5	Perturbative Account of Electron Correlation Effects in the Internal Rotational Barrier of Molecules: A State Specific Strategy *Sudip Chattopadhyay*	155
Index		183

PREFACE

In An Introduction to Electronic Structure Theory, Quantum Information Theory is applied to donor-acceptor systems. Reaction stages and charge-transfer phenomena are described, continuities of probability and phase distributions are explored, and resultant information descriptors combining classical and nonclassical contributions are summarized.

The authors describe the most efficient method for studying the electronic structure of solids, the magnetic dilution method, or the study of the magnetic susceptibility of diluted solid solutions of paramagnetic oxides in diamagnetic isomorphous matrices.

A review of the mathematical modeling and investigation of the electronic structure of some nanomaterials, composite materials, and graphene is presented using the Parameterized Model number 3 (PM3) semi-empirical method.

A basic introduction of electronic structure theory with commonly used notation is provided, as well as its applications for studying the physical properties of materials.

Lastly, based on a concept of "different prescription for different correlation", a multireference Brillouin-Wigner perturbation scheme with improved virtual orbitals is presented as an accurate and affordable computational protocol for treating electronic states plagued by quasidegeneracy.

Chapter 1 - Quantum Information Theory (QIT) is applied to donor-acceptor systems. Reaction stages and charge-transfer phenomena are described, continuities of probability and phase distributions are explored, and resultant information descriptors combining the classical (modulus/probability) and nonclassical (phase/current) contributions are summarized. The net production of resultant gradient information is shown to be generated by a nonvanishing phase-source. A novel integral concept of information content is proposed, the density of which obeys the classical relation between distributions of the gradient (Fisher) and global (Shannon) measures. The grand-ensemble description of open systems is summarized, the ensemble-average descriptors of the mutually-open fragments of externally-closed reactive system are determined, and the phase distinction between their bonded and nonbonded status is examined. Additive QIT descriptors of reactants are introduced and subsystem *phase*-equalization is established. The reactive systems, *multi*-component configurations of equidensity orbitals, and *one*-dimensional model of a gradual opening of subsystems are explored as illustrative examples.

Chapter 2 - The chapter is devoted to the description of the most efficient method for studying the electronic structure of solids – the magnetic dilution method, e.g., the study of magnetic susceptibility of diluted solid solution of paramagnetic oxides in diamagnetic isomorphous matrices. The method gives the possibility to determine the electronic state of a single paramagnetic atom, the energetics of magnetic exchange between paramagnetic atoms, and the character of paramagnetic atoms distribution in a diamagnetic matrix. The influence of the composition of a diamagnetic matrix can be traced and explained. Complex oxides with perovskite, layered K_2NiF_4 type, and spinel structure are considered as the examples of the magnetic dilution method application.

Chapter 3 - This review has been devoted to the mathematical modeling and investigation of the electronic structure of some nanomaterials, composite materials, and graphene by using the Parameterized Model number 3 (PM3) semi-empirical method. One of the variants of the molecular orbitals method - the semi-empirical Wolfsberg–Helmholz method was used to investigate the properties of the nanoparticles. For construction of molecular orbitals Ag_{16} are used 5s-, 5py-, 5pz-, and 5px- valence Slater atomic orbitals of silver atoms. The analytic expression of the basis Slater functions was defined. The orbital energies, ionization potential, the total electronic energy, and effective charge of atoms of silver nanoparticles were calculated by the solution of equations of the molecular orbitals method. The possibility of using the simple computer program developed in Delphi Studio working undo MS Windows OS for carrying out the quantum mechanical calculation of the electronic structure of nanoparticles has been investigated. The theoretical methodology is described for the realization of this simple computer. The numerous quantum mechanical calculations show that this computer program works correctly and it is useful for use based on Slater Atomic Orbitals. The electronic structure of the gold nanoparticles was investigated by the semi-empirical Wolfsberg – Helmholz method. As the atomic orbitals used 6s-, $6p_y$-, $6p_z$- and $6p_x$- Slater atomic orbitals of gold atoms. The orbital energies, potential ionization, the total electronic energy, and effective charge of atoms of gold nanoparticles were calculated. The results indicate that the gold nanoparticles are soft, electrophile, and conductive material. Theoretical models of shockproof composite materials based on two-layer graphene and multilayer polyvinylidene fluoride $C_{124}H_{40} + n(H-(C_2H_2F_2)_5-H)$ (n = 1,2,...,8) are constructed. The electronic structure is studied using the semiempirical PM3 method that is one of the options of the molecular orbital method. The orbital energies, ionization potentials, total electron energies, strength, and other properties of the considered material are calculated based on the theoretical models. The outlooks for application of these materials in the military field for manufacturing super strong and lightweight flak jackets are considered.

Chapter 4 - Recently, the electronic structure theory plays an important role for studying the various physical properties of materials through ab-initio approach. It depends equally upon density functional theory (DFT) and wave function theory (WFT). The purpose of this chapter is to give a basic introduction of electronic structure theory with commonly used notation and its applications for studying the physical properties of materials.

Chapter 5 - Based on a concept of "different prescription for different correlation", a multireference Brillouin–Wigner perturbation scheme with improved virtual orbitals (IVO-BWMRPT) has been presented as an accurate and affordable computational protocol for treating the electronic states plagued by quasidegeneracy. It deals with only a single-root problem in the Hilbert space treating all components of the model space on the same footing. The IVO-BWMRPT approach has several attractive properties (explicit size-extensivity and intruder-free nature without using any numerical threshold or ad hoc parameter) for investigating various chemical processes and problems. The IVOs were generated using the complete active space configuration interaction (CASCI) scheme and exploited as a means of recognizing state-specific nondynamical (neardegeneracy) effects. In IVO-BWMRPT, given an IVO-CASCI wave function, the re maining dynamical correlation can be efficiently recovered using BWMRPT scheme. Main motivation in using the IVO-CASCI is that it does not need iterations, nor does it face convergence difficulties that may occur in complete active space self-consistent field (CASSCF) estimations, although IVO-CASCI function cherishes the appealing trait of the CASSCF wave function. Investigations of the torsion of diimide and hydrazine demonstrate that biradicaloid electronic structure can be described nicely with this method where the usual single-reference description fails. As the present approach does not utilize any parameter or numerically unstable operation, the surfaces provided by IVO-BWMRPT are smooth all along the reaction path. A promising accordance between the present estimates and literature values has been found. The IVO-based perturbative technique can open the possibility for an accurate description of the energy surfaces for the ground and excited states of small to large systems at the ab initio level in a simplified fashion.

In: An Introduction to Electronic Structure ... ISBN: 978-1-53618-411-2
Editor: Nadia T. Paulsen © 2020 Nova Science Publishers, Inc.

Chapter 1

RESULTANT INFORMATION APPROACH TO DONOR-ACCEPTOR SYSTEMS[†]

Roman F. Nalewajski[*]
Department of Theoretical Chemistry,
Jagiellonian University, Cracow, Poland

ABSTRACT

Quantum Information Theory (QIT) is applied to donor-acceptor systems. Reaction stages and charge-transfer phenomena are described, continuities of probability and phase distributions are explored, and resultant information descriptors combining the classical (modulus/probability) and nonclassical (phase/current) contributions are summarized. The net production of resultant gradient information is shown to be generated by a nonvanishing phase-source. A novel integral

---

[†] The following notation is adopted: A denotes a *scalar*, $\mathbf{A}$ is the row or column *vector*, $\mathbf{A}$ represents a square (or rectangular) *matrix*, and the dashed symbol $\hat{A}$ stands for the quantum-mechanical *operator* of the physical property A. The logarithm of Shannon's information measure is taken to an arbitrary but fixed base: $\log = \log_2$ corresponds to the information content measured in *bits* (binary digits), while $\log = \ln$ expresses the amount of information in *nats* (natural units): 1 nat = 1.44 bits.

[*] Corresponding Author's Email: nalewajs@chemia.uj.edu.pl.

concept of information content is proposed, the density of which obeys the classical relation between distributions of the gradient (Fisher) and global (Shannon) measures. The grand-ensemble description of open systems is summarized, the ensemble-average descriptors of the mutually-open fragments of externally-closed reactive system are determined, and the phase distinction between their bonded and nonbonded status is examined. Additive QIT descriptors of reactants are introduced and subsystem *phase*-equalization is established. The reactive systems, *multi*-component configurations of equidensity orbitals, and *one*-dimensional model of a gradual opening of subsystems are explored as illustrative examples.

Keywords: chemical reactivity theory, Continuity relations, Donor-Acceptor Systems, Information theory, Phase-Equalization, Resultant entropy/information

1. INTRODUCTION

The classical Information Theory (IT) [1-8] of Fisher [1] and Shannon [3] has already been successfully applied in entropic interpretations of molecular electron densities or probability distributions [9-12]. Information principles have been explored [9-16] and density pieces attributed to Atoms-in-Molecules (AIM) have been approached [12, 16-20] providing the IT basis for the intuitive stockholder division of Hirshfeld [21]. Patterns of chemical bonds have been extracted from electronic communications in molecules [9-11, 22-32], information distributions in molecules have been explored [9-11, 33, 34], and the nonadditive Fisher information [9-11, 35, 36] has been linked to the Electron Localization Function (ELF) [37-39] of modern Density Functional Theory (DFT) [40-45]. This analysis has also formulated the Contragradience (CG) probe for localizing chemical bonds [9-11, 46], and Orbital Communication Theory (OCT) of the chemical bond [9-11, 22-32] has identified bridge interactions of AIM through intermediate orbitals [11, 47-52], realized via the cascade propagations in molecular information systems.

In molecular Quantum Mechanics (QM) the complex nature of electronic states calls for an extension of the classical (probability)

concepts of the entropy and information [53-62]. The wavefunction phase or its gradient determining the effective velocity of probability density, give rise to nonclassical supplements of the classical IT measures of Fisher [1] and Shannon [3], of the information/entropy contained in electronic probability distributions. The resultant IT descriptors of electronic states combine the *classical* (modulus/probability) and *nonclassical* (phase/current) contributions. The state overall gradient-information is then proportional to the expectation value of the system kinetic energy of electrons. This generalized treatment allows one to interpret the variational principle for electronic energy as equivalent information rule, and to use the molecular virial theorem [63] in general reactivity considerations [64-67]. Such combined descriptors are also required for the phase distinction between the bonded (entangled) and nonbonded (disentangled) states of molecular subsystems, for the same set of the fragment electron densities [68, 69], e.g., the substrate fragments of reactive systems.

The extremum principles for the global and local (gradient) measures of the state resultant entropy have determined the phase-transformed (equilibrium) states, identified by the probability-dependent "thermodynamic" phases [53-58, 60-69]. The minimum-energy principle of QM has been recently interpreted [64-67] as physically-equivalent variational rule for the resultant gradient information, proportional to the state average kinetic energy. In the *grand*-ensemble they both determine the same thermodynamic equilibrium in an externally-open molecular system. This equivalence parallels identical predictions resulting from the minimum-energy and maximum-entropy principles in ordinary thermodynamics [70].

We begin the present analysis with a brief summary of conventional stages of chemical reactions in the donor-acceptor systems and of admissible electronic communications they imply. The continuity relations for the modulus (probability) and phase (current) distributions of electronic states implied by the molecular Schrödinger equation (SE) will be explored, and an overview of the resultant entropy/information descriptors will be given. A nonclassical origin of the overall information production will be emphasized and the novel concept of a *geometric* measure of the combined entropy (uncertainty) content will be introduced, which obeys the classical relation between densities-per-electron of the classical IT measures: global of Shannon and local of Fisher. The grand-ensemble description of open reactive systems and their fragments will be examined, the phase distinction between states of the bonded (entangled) and non-bonded (disentangled) subsystems will be explored, and the additive IT descriptors will be used to demonstrate the phase equalization in mutually-open fragments. The closed and open states of reactants and *multi*-component systems of the occupied equidensity orbitals (EO), as well as the model *one*-dimensional problem of the opaque division wall will be explored as illustrative examples.

2. REACTION STAGES IN DONOR-ACCEPTOR SYSTEMS

The QM and IT provide a solid basis for determining and understanding the electronic structure of molecules, and for explaining general trends in their chemical behavior. Qualitative considerations on preferences in chemical reactions usually emphasize changes in reactant energies induced by displacements (perturbations), real or hypothetical, in parameters describing electronic states at conventional reaction stages [9, 11, 62, 71-78].

In reactivity theory one thus examines displacements in parameters determining the system electronic Hamiltonian, and determines responses to such perturbations [71, 72]. These global and local degrees-of-freedom of molecular Hamiltonians in the Born-Oppenheimer (BO) approximation include the system overal number of electrons N and the external potential $v(r)$ generated by the rigid nuclear frame:

$$\hat{H}(N,v) = \sum_{i=1}^{N} v(r_i) + [\hat{T}(N) + \hat{U}(N)] \equiv \hat{V}(N,v) + \hat{F}(N) \qquad (1)$$

Here, $\hat{V}(N,v)$ stands for the quantum-mechanical operator of the electron attraction by the nuclei in their fixed positions Q, which generate the external potential $v(r; Q) \equiv v(r)$ for the system electrons, and $\hat{F}(N)$ denotes the sum of remaining kinetic $[\hat{T}(N)]$ and electron-repulsion $[\hat{U}(N)]$ terms.

Consider the expectation value of the system electronic energy in the molecular ground-state $\Psi[N, v] = \Psi[\rho]$ yielding electron density $\rho[N, v; r] \equiv \rho(r)$,

$$E[N, v] = \langle \Psi[N, v] | \hat{H}(N, v) | \Psi[N, v] \rangle$$
$$= \int v(r) \rho(r) \, dr + \langle \Psi[N, v] | \hat{F}(N) | \Psi[N, v] \rangle$$
$$\equiv V[\rho] + F[\rho] \equiv E_v[\rho], \qquad (2)$$

the lowest eigenvalue $E[N, v]$ of electronic Hamiltonian,

$$\hat{H}(N, v) \Psi[N, v] = E[N, v] \Psi[N, v], \qquad (3)$$

given also by the corresponding density functional $E_v[\rho]$ [40-45]. The mutually or externally open reactants are free to exchange electrons with the hypothetical (macroscopic) reservoir. Their average number of electrons is thus a continuous variable justifying the use of populational derivatives of the system energy [71-85] or the resultant gradient information [64-66] in describing the charge-transfer (CT) phenomena in reactive systems consisting of the electron donor (basic) and acceptor (acidic) substrates.

In reactivity considerations one thus explores infinitesimal displacements $\{dN, \delta v(r)\}$ of the system parameters, with dN representing a change of the average number of electrons in the *grand*-ensemble [79, 80], and examines responses to such elementary perturbations. The overall state-parameters of the geometrically "frozen" subsystems $\alpha = \{A(acid)$ and $B(base)\}$ in the reactive system $R = A\text{----}B$ are given by corresponding sums of the associated descriptors of the separate fragments,

$$N_R = \sum_\alpha N_\alpha \equiv N \quad \text{and} \quad v_R(r) = \sum_\alpha v_\alpha(r) \equiv v(r). \qquad (4)$$

They can be thus classified as "extensive" in character. Accordingly, their derivative energy conjugates, the system global chemical potential

$$\mu[N, v] = \partial E[N, v]/\partial N = \mu \qquad (5)$$

and its electron density

$$\rho[N, v; r] = \delta E[N, v]/\delta v(r) = \rho(r), \qquad (6)$$

should be regarded as the associated "intensities." These energy conjugates are thus determined by the corresponding derivatives of the ensemble-average electronic energy, the functional of the system average "extensive" parameters of state. Similar relations link the ground-state descriptors of the separate (isolated) subsystems $\alpha^0 \in \{A^0, B^0\}$:

$$\mu_\alpha^0[N_\alpha^0, v_\alpha] = \partial E[N_\alpha, v_\alpha]/\partial N_\alpha|_0 \quad \text{and}$$

$$\rho_\alpha^0[N_\alpha^0, v_\alpha; r] = \delta E[N_\alpha^0, v_\alpha]/\delta v_\alpha(r)|_0. \qquad (7)$$

They also confirm the "intensive" character of the energy-conjugates $\{\mu_\alpha^0, \rho_\alpha^0\}$ of the "extensive" state variables $\{N_\alpha^0, v_\alpha\}$ in electronic Hamiltonians of subsystems.

In reactivity considerations one customarily recognizes several hypothetical stages of chemical processes involving either the mutually-closed (nonbonded, "disentangled") or mutually-open (bonded, "entangled") fragments (see Figure 1), e.g., the substrate or product subsystems in a bimolecular chemical reaction [71, 72].

Consider the substrates $\alpha = (A, B)$ in the specified donor-acceptor reactive system R = A----B. The isolated (separated) subsystems exhibit different global levels of the internally equalized chemical potentials, $\mu_A^0 < \mu_B^0$,

$$\{\mu_\alpha^0(r) \equiv \delta E_\alpha[\rho_\alpha^0]/\delta \rho_\alpha^0(r) = \mu_\alpha[\rho_\alpha^0, v_\alpha] = \mu_\alpha^0 \equiv \partial E(N_\alpha, v_\alpha)/\partial N_\alpha|_0\}, \qquad (8)$$

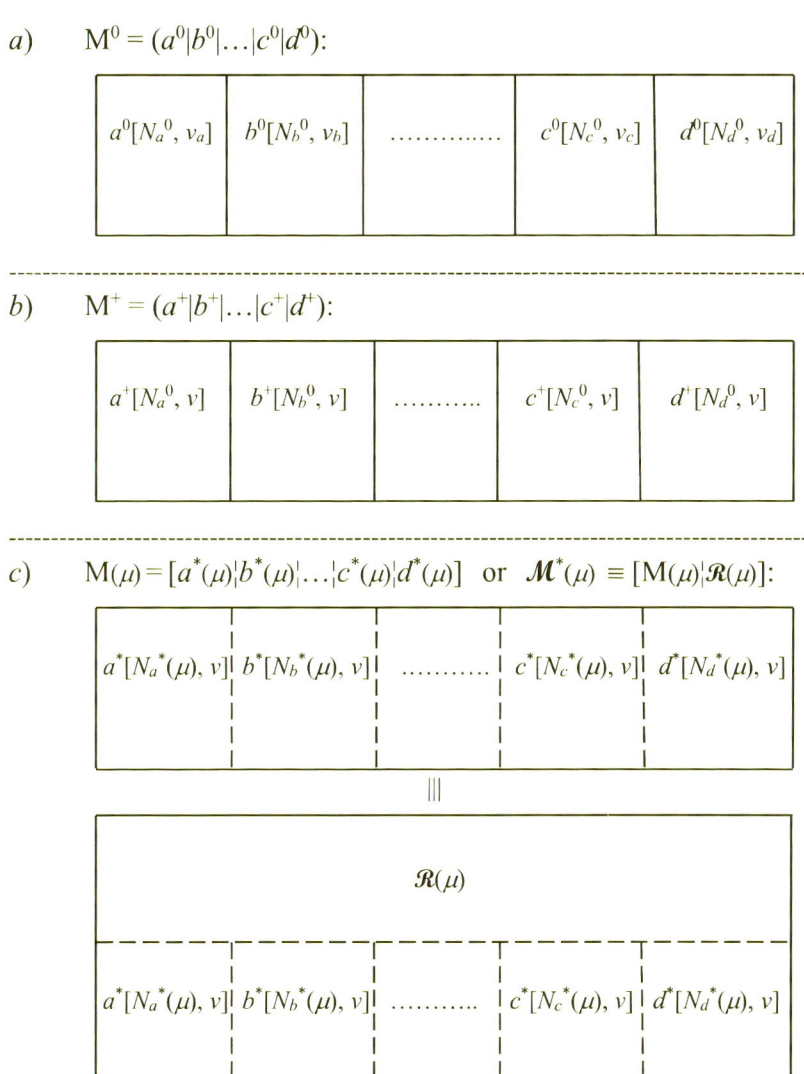

Figure 1. Hypothetical reaction stages in relaxing the electronic structure of molecular fragments $\alpha \in \{a, b, ..., c, d\}$ from the isolated species $\{\alpha^0\}$ in the promolecule $M^0 = (a^0|b^0| ... |c^0|d^0)$ (Panel a), via the polarized subsystems $\{\alpha^+\}$ in $M^+ = (a^+|b^+| ... |c^+|d^+)$ (Panel b), to the equilibrium (bonded) fragments $\{\alpha^*(\mu)\}$ in $M(\mu) = [a^*(\mu)|b^*(\mu)| ... |c^*(\mu)|d^*(\mu)]$ (Panel c) for the molecular chemical potential $\mu = \mu_M$. The latter can be equivalently viewed as molecular (internally open) composite molecular system $M(\mu)$ coupled to the macroscopic electron reservoir $\mathcal{R}(\mu)$ in $\mathcal{M}^*(\mu) = [M(\mu)|\mathcal{R}(\mu)]$.

a) $R^o(\mu_A, \mu_B) = [\mathcal{R}_A(\mu_A) \vdots A^o(\mu_A) | B^o(\mu_B) \vdots \mathcal{R}_B(\mu_B)]$:

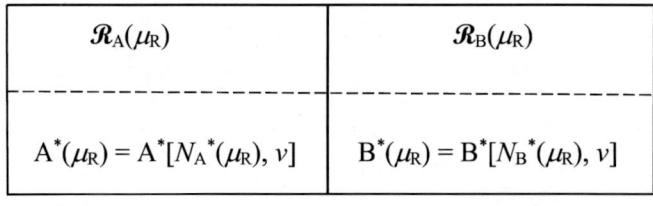

b) $R^*(\mu_R) = [\mathcal{R}_A(\mu_R) \vdots A^*(\mu_R) | B^*(\mu_R) \vdots \mathcal{R}_B(\mu_R)] \equiv [\mathcal{R}_R(\mu_R) \vdots A^*(\mu_R) \vdots B^*(\mu_R)]$:

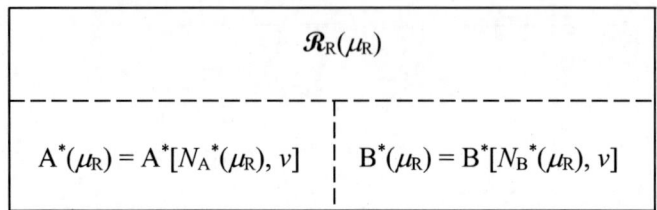

Figure 2. Externally open reactants of donor-acceptor systems $R^o(\mu_A, \mu_B)$ (Panel *a*) and $R^*(\mu_R)$ (Panel *b*) involving the separate and common electron reservoirs of subsystems, respectively. Panel *b* additionally illustrates that equalization of chemical potentials of the separate reservoirs, $\mu_A = \mu_B = \mu_R$, effectively opens both reactants mutually.

where $E_\alpha[\rho_\alpha^0[v_\alpha]] \equiv E[N_\alpha^0, v_\alpha]$ denotes the ground-state energy of α^0. Their nonbonded status, when they conserve the overall (integer) numbers of electrons $\{N_\alpha = N_\alpha^0\}$ of the separated reactants $\{\alpha^0\}$, implies the vanishing exchanges of electrons with the reaction partner, i.e., their *mutual-closeness* symbolized by the *solid* vertical line separating the two subsystems, or with the macroscopic reservoir (*external*-closeness) in the ensemble representation. For example, the initial, *promolecular* stage $R^0 \equiv (A^0|B^0)$ signifies the "frozen" ground-state densities $\{\rho[N_\alpha^0, v_\alpha] = \rho_\alpha^0 \equiv N_\alpha^0 p_\alpha^0\}$ of isolated fragments $\{\alpha^0\}$ shifted to their "molecular" positions in R, for the specified mutual separation and orientation of these geometrically rigid fragments in the reactive system; here $p_\alpha^0(r)$ stands for the fragment probability distribution: $\int p_\alpha^0(r)\, dr = 1$. This initial (reference) stage of a chemical reaction thus generates a nonstationary "molecular" distribution of electrons, nonequilibrium for the *molecular* external potential $v = v_A + v_B$,

$$\rho^0 = \rho_A^0 + \rho_B^0 \neq \rho[N_R, v], \tag{9}$$

different from the equilibrium (ground-state) density $\rho[N_R, v]$ for the overall electron number $N_R = N_A + N_B$ and the *resultant* external potential v. This promolecular state thus implies nonequalized local chemical potentials in both subsystems, determined by the functional derivatives $\{\mu_\alpha(r) \equiv \delta E[\{\rho_\gamma\}]/\delta\rho_\alpha(r)\}$ of the density bifunctional for the electronic energy in R, $E[\rho_A, \rho_B; v]$, calculated for the "frozen" densities $\{\rho_\alpha^0\}$ of isolated subsystems:

$$\mu_A[v; r] \neq \mu_A[v; r'] \ldots \quad \text{and} \quad \mu_B[v; r] \neq \mu_B[v; r'] \ldots \tag{10}$$

The intermediate, *polarized* reactive system $R^+ \equiv (A^+|B^+)$ similarly combines densities $\{\rho_\alpha^+ = \rho_\alpha[N_A^0, N_B^0; v] = N_\alpha^+ p_\alpha^+\}$ of the equilibrium (mutually-closed) subsystems $\{\alpha^+\}$, separated above by the *solid* vertical line symbolizing $\{N_\alpha^+ = N_\alpha^0\}$, thus exhibiting different levels of their (internally-equalized) global chemical potentials:

$$\mu_A^+ \equiv \mu_A[\rho_A^+, \rho_B^+; v] = \mu_A^+[v; r] = \mu_A^+[v; r'] = \ldots$$
$$< \mu_B^+ \equiv \mu_B[\rho_A^+, \rho_B^+; v] = \mu_B^+[v; r] = \mu_B^+[v; r'] = \ldots \tag{11}$$

The implicit barrier for the flow of electrons between polarized reactants is eventually lifted in the final *equilibrium* stage of a "molecular" reactive system, $R^* = (A^*|B^*) \equiv R$, after the optimum $B^+ \to A^+$ CT between the initially polarized reactants (Figure 2). This freedom to exchange electrons has been symbolized above by the *broken* line separating the bonded subsystems $\{\alpha^*\}$ which exhibit the ensemble-average densities

$$\{\rho_\alpha^*(r) \equiv N_\alpha^* p_\alpha^*(r) = \rho_\alpha[N_A^*, N_B^*; v; r] \equiv \rho_\alpha[\rho_R; r],$$

$$\sum_\alpha \rho_\alpha^*(r) = \rho_R(r), \qquad N_\alpha^* = \int \rho_\alpha^*(r)\, dr = N_\alpha[N_R; v]. \tag{12}$$

For the lack of the separating boundary in R^* its mutually-open subsystems must explore the electron probability distribution of the whole reactive system:

$$p_A^*(r) \equiv \rho_A^*(r)/N_A^* = p_B^*(r) \equiv \rho_B^*(r)/N_B^* = p_R(r) = \rho_R(r)/N_R. \tag{13}$$

Indeed, the density additivity implies

$$\rho_R(r) = \rho[N_R, v; r] \equiv N_R\, p_R(r) = (N_A^* + N_B^*)\, p_R(r)$$
$$= \rho_A^*(r) + \rho_B^*(r) = N_A^* p_A^*(r) + N_B^* p_B^*(r)$$

and hence:

$$p_\alpha^*(r) = p_R(r) \quad \text{and} \quad \rho_\alpha^*(r)/\rho_R(r) = N_\alpha^*/N_R \equiv P_\alpha^*, \quad \sum_\alpha P_\alpha^* = 1.$$

The effective mutual-openess signifies the bonded character of reaction partners and hence also the equalization of their chemical potentials at the "molecular" level, in R^* as a whole,

$$\mu_\alpha^* \equiv \mu_\alpha[\rho_A^*, \rho_B^*; v] \equiv \mu_\alpha[N_R; v] = \mu[N_R, v] \equiv \mu_R[\rho_R, v], \quad \alpha = A, B. \tag{14}$$

All these stages are assumed to be globally isoelectronic:

$$N_R \equiv N_A{}^0 + N_B{}^0 \equiv N_R{}^0 = N_A{}^+ + N_B{}^+ \equiv N_R{}^+ = N_A{}^* + N_B{}^* \equiv N_R.$$

The bonded fragments generally exhibit fractional values of the average numbers of electrons, due to a finite amount $N_{CT} > 0$ of the inter-reactant CT,

$$N_A{}^* = \int \rho_A{}^* d\boldsymbol{r} = N_A{}^0 + N_{CT} > N_A{}^0,$$

$$N_B{}^* = \int \rho_B{}^* d\boldsymbol{r} = N_A{}^0 - N_{CT} < N_B{}^0. \tag{15}$$

Such mutually-open fragments call for an ensemble description of their average densities and electron populations. Their identity is properly defined only when they represent the mutually-closed fragments of the macroscopic subsystems composed a the reactant fragments coupled to their own (separate) electron reservoirs [64-67]. Therefore, the substrate identity represents a meaningful concept only when both subsystems remain mutually closed.

The fragment properties in the final, equilibrium reactive system R combining the bonded fragments can indeed be inferred only indirectly, by externally opening the mutually closed subsystems of R$^+$, with respect to their separate (macroscopic) electron reservoirs $\{\mathcal{R}_\alpha{}^+\}$ in the polarized composite system of Figure 2a:

$$\mathcal{M}_R{}^+ = [\mathcal{R}_A{}^+(\mu_A{}^+)|A^+(\mu_A{}^+)|B^+(\mu_B{}^+)|\mathcal{R}_B{}^+(\mu_B{}^+)] \equiv [\mathcal{M}_A{}^+(\mu_A{}^+)|\mathcal{M}_B{}^+(\mu_B{}^+)]. \tag{16}$$

The chemical potentials of such mutually-closed subsystems can be subsequently equalized at the "molecular" level,

$$\mu_A{}^+ \equiv \mu_A{}^* = \mu_B{}^+ \equiv \mu_B{}^* = \mu_R[N_R, v], \tag{17}$$

of the chemical potential of R as a whole. Such a composite system $\mathcal{M}_R{}^*$ effectively involving a common electron reservoir $\mathcal{R}_A{}^*(\mu_R) = \mathcal{R}_B{}^*(\mu_R) = \mathcal{R}^*(\mu_R) \equiv \mathcal{R}$ (Figure 2b),

$$\mathcal{M}_R^* = [\mathcal{R}_A^*(\mu_R) | A^*(\mu_R) | B^*(\mu_R) | \mathcal{R}_B^*(\mu_R)] \equiv [\mathcal{M}_A^*(\mu_R) | \mathcal{M}_B^*(\mu_R)]$$
$$= [\mathcal{R}^*(\mu_R) | A^*(\mu_R) | B^*(\mu_R)] \equiv [\mathcal{R}(\mu_R) | R^*(\mu_R)] \qquad (18)$$

then combines the formally bonded reactant subsystems. Indeed, the substrate chemical potentials equalized at the molecular level in both subsystems, when $\{\mathcal{R}_\alpha^+(\mu_R) \equiv \mathcal{R}_\alpha^*(\mu_R) = \mathcal{R}(\mu_R)\}$, effectively describe a single "molecular" reservoir coupled to the whole reactive system R: $[\mathcal{R}|A^*|B^*] = (\mathcal{R}^*|R^*)$. The open character of each subsystem implies that each fragment effectively exhausts the whole electron probability density $p_R = \rho_R/N_R$ in $R = R^*(\mu_R)$ [see Eq. (13)] so that electron densities of the mutually-open substrates,

$$\rho_\alpha^*(r) = N_\alpha^* p_R(r) = (N_\alpha^*/N_R) \rho_R(r) = P_\alpha^* \rho_R(r), \qquad \alpha = A, B, \qquad (19)$$

where $P_\alpha^* = N_\alpha^*/N_R$ denotes the global probability of α^* in R, reconstruct the equilibrium electron density in the whole system:

$$\sum_\alpha \rho_\alpha^*(r) = \rho_R(r) (\sum_\alpha P_\alpha^*) = \rho_R(r).$$

The "frozen" electron configuration of both substrates in the promolecular system

$$R^0 = (A^0 | B^0) = ([a_1^0 | a_2^0 | ...] | [b_1^0 | b_2^0 | ...])$$

precludes all electron communications [9-11, 22-32, 62] between constituent fragments/states of the "frozen" acidic $A^0 = [a_1^0 | a_2^0 | ...]$ and basic $B^0 = [b_1^0 | b_2^0 | ...]$ reactants. The polarized acidic $A^+ = [a_1^{+!} | a_2^{+!} | ...]$ and basic $B^+ = [b_1^{+!} | b_2^{+!} | ...]$ subsystems in

$$R^+ = (A^+ | B^+) = ([a_1^{+!} | a_2^{+!} | ...] | [b_1^{+!} | b_2^{+!} | ...])$$

communicate only internally, $\{a_i^+ \to a_j^+\}$ and $\{b_k^+ \to b_l^+\}$, while the mutual "opening" (bonding, entangling) of the equilibrium molecular fragments $A^* = [a_1^{*}|a_2^{*}|...]$ and $B^* = [b_1^{*}|b_2^{*}|...]$ in

$$R^* = (A^*|B^*) = ([a_1^{*}|a_2^{*}|...]|[b_1^{*}|b_2^{*}|...])$$

implies that two complementary subsystems can communicate both internally and externally, between themselves, $\{a_i^* \to b_k^*\}$, with each propagation link contributing to the resultant pattern of entropic bond-orders [9-11, 22-32, 62]. Therefore, the above reaction stages also reflect a hierarchy of the allowed communications between the fragment constituent AIM or the basis functions they contribute to chemical bonds in the reactive system.

3. CONTINUITY EQUATIONS

The dynamics of quantum electronic states is determined by the Schrödinger equation (SE) of molecular QM, which also determines time evolutions of the state modulus and phase components, summarized by the relevant continuity equations. For simplicity let us consider a single electron in state $|\psi(t)\rangle$ at time t, or the associated (complex) wavefunction in position representation:

$$\psi(r, t) = \langle r|\psi(t)\rangle = R(r, t) \exp[i\phi(r, t)], \quad \phi(r, t) \geq 0. \tag{20}$$

Its modulus (R) and phase (ϕ) components determine the state physical distributions of electron probability and current densities:

$$p(r, t) = \psi(r, t)^* \psi(r, t) = R(r, t)^2, \tag{21}$$

$$\begin{aligned} j(r, t) &= [\hbar/(2mi)] [\psi(r, t)^* \nabla \psi(r, t) - \psi(r, t) \nabla \psi(r, t)^*] \\ &= (\hbar/m) p(r, t) \nabla \phi(r, t) \equiv p(r, t) V(r, t). \end{aligned} \tag{22}$$

The effective velocity $V(r, t)$ of probability "fluid" measures the current-per-particle and reflects the state phase-gradient:

$$V(r, t) = j(r, t)/p(r, t) = (\hbar/m) \nabla \phi(r, t) = dr(t)/dt.$$

In molecular scenario the electron is moving in external potential $v(r)$ due to the "frozen" nuclear frame of BO approximation. The electronic Hamiltonian

$$\hat{H}(r) = -(\hbar^2/2m)\nabla^2 + v(r) \equiv \hat{T}(r) + v(r), \qquad (23)$$

where $\hat{T}(r)$ denotes its kinetic part, determines the quantum dynamics of electronic states as expressed by SE:

$$i\hbar [\partial \psi(r, t)/\partial t] = \hat{H}(r) \psi(r, t). \qquad (24)$$

The associated temporal evolutions of the modulus $R(r, t)$ and phase $\phi(r, t)$ functions ultimately generate the continuity equations for instantaneous electronic probability $p(r, t)$ and current $j(r, t)$ distributions.

The total time-derivative of $p(r, t) = p[r(t), t]$,

$$\sigma_p(r, t) \equiv dp(r, t)/dt = \partial p[r(t), t]/\partial t + [dr(t)/dt] \cdot \partial p[r(t), t]/\partial r$$
$$= \partial p(r, t)/\partial t + V(r, t) \cdot \nabla p(r, t), \qquad (25)$$

defines the local probability "source" $\sigma_p(r, t)$. It measures the time rate of change in an infinitesimal volume element of the probability fluid moving with the velocity $V(r, t) = dr(t)/dt$, while the partial derivative $\partial p(r, t)/\partial t$ refers to the volume element around the fixed point in space. This probability-dynamics expresses the continuity relation

$$\sigma_p(r, t) = \partial p(r, t)/\partial t + \nabla \cdot j(r, t) = 0,$$

$$\nabla \cdot j(r, t) = V(r, t) \cdot \nabla p(r, t) + p(r, t) \nabla \cdot V(r, t)$$

$$= (\hbar/m) [\nabla \phi(r,t) \cdot \nabla p(r,t) + p(r,t) \nabla^2 \phi(r,t)], \qquad (26)$$

since Eq. (25) implies the vanishing divergence of velocity field, related to the phase-Laplacian $\nabla^2 \phi(r,t) = \Delta\phi(r,t)$,

$$\nabla \cdot V(r,t) = (\hbar/m) \Delta\phi(r,t) = 0 \quad \text{or} \quad \Delta\phi(r,t) = 0. \qquad (27)$$

The associated dynamical equation for the state modulus component reads

$$\partial R(r,t)/\partial t = -V(r,t) \cdot \nabla R(r,t), \qquad (28)$$

while the phase-dynamics predicts:

$$\partial \phi(r,t)/\partial t = [\hbar/(2m)] \{R(r,t)^{-1} \Delta R(r,t) - [\nabla\phi(r,t)]^2\} - v(r)/\hbar. \qquad (29)$$

The preceding equation can be also interpreted as the phase-continuity relation when one realizes that the effective velocity $V(r,t)$ of the probability-current $j(r,t) = p(r,t) V(r,t)$ also determines the phase-flux

$$J(r,t) = \phi(r,t) V(r,t)$$

and its divergence [see Eq. (27)]:

$$\nabla \cdot J(r,t) = \nabla\phi(r,t) \cdot V(r,t) = (\hbar/m) [\nabla\phi(r,t)]^2.$$

The above phase-flow descriptor then generates a finite source of this wavefunction component:

$$\sigma_\phi(r,t) \equiv d\phi(r,t)/dt = \partial\phi(r,t)/\partial t + \nabla \cdot J(r,t)$$
$$= \partial\phi(r,t)/\partial t + V(r,t) \cdot \nabla\phi(r,t) \neq 0. \qquad (30)$$

Using Eq. (29) finally gives the following local phase-production:

$$\sigma_\phi(r,t) = [\hbar/(2m)]\{R(r,t)^{-1}\Delta R(r,t) + [\nabla\phi(r,t)]^2\} - v(r)/\hbar. \tag{31}$$

To summarize, the effective velocity of probability-current also determines the phase-flux in molecules. The source (net production) of the classical probability-variable of electronic states identically vanishes, while that of their nonclassical, phase-component, determined by both state components and the external potential, remains finite.

For the given time $t = t_0$ the probability density $p(r, t_0) \equiv p(r)$ reflects the (statical) structure "of being" contained in the electronic state in question, while the current density $j(r) = (\hbar/m)p(r)\nabla\phi(r)$ determines its (dynamical) content of the structure "of becoming" [86]. They both contribute to the resultant measures of the overall information or entropy in quantum electronic states. We have argued above that temporal evolutions of these two aspects of the static electronic structure are determined by their separate continuity relations:

$$\partial p(r)/\partial t = -\nabla \cdot j(r) = -(\hbar/m)\nabla p(r)\cdot\nabla\phi(r) \quad \text{and}$$

$$\partial\phi(r)/\partial t = -\nabla\cdot J(r) + \sigma_\phi(r) = -(\hbar/m)[\nabla\phi(r)]^2 + \sigma_\phi(r)$$
$$= [\hbar/(2m)]\{\Delta p(r)/[2p(r)] - \tfrac{1}{4}[\nabla\ln p(r)]^2 - [\nabla\phi(r)]^2\} - v(r)/\hbar. \tag{32}$$

These partial derivatives, which reflect time-rates at the specified location in space, explicitly confirm that both the probability and (phase/current)-components influence the time-rates of these quantum physical distributions.

4. RESULTANT INFORMATION/ENTROPY MEASURES

The Fisher's measure of the gradient information contained in probability density is reminiscent of von Weizsäcker's [87] inhomogeneity correction in the density functional for electronic kinetic energy:

Resultant Information Approach to Donor-Acceptor Systems 17

$$I[p;t] = \int [\nabla p(\mathbf{r},t)]^2/p(\mathbf{r},t)\, d\mathbf{r} = 4\int [\nabla R(\mathbf{r},t)]^2\, d\mathbf{r} \equiv I[R;t]$$
$$= \int p(\mathbf{r},t)\, [\nabla \ln p(\mathbf{r},t)]^2\, d\mathbf{r} \equiv \int p(\mathbf{r},t)\, I_p(\mathbf{r},t)\, d\mathbf{r}. \qquad (33)$$

This classical measure characterizes an effective "narrowness" of $p(\mathbf{r}, t)$, a degree of the particle position-determinicity. It represents the IT descriptor complementary to Shannon's global entropy

$$S[p;t] = -\int p(\mathbf{r},t) \ln p(\mathbf{r},t)\, d\mathbf{r} = -2\int R(\mathbf{r},t)^2 \ln R(\mathbf{r},t)\, d\mathbf{r} \equiv S[R;t], \qquad (34)$$

which reflects a "spread" of probability density, i.e., the particle position-indeterminicity.

These classical IT measures generalize naturally into the corresponding resultant descriptors combining contributions from both the modulus (probability) and phase (current) components [53-62]. For example, the overall gradient-information in state $|\psi(t)\rangle$ is given by the expectation value of the (Hermitian) information operator $\hat{I}(\mathbf{r})$ [35], related to kinetic-energy operator $\hat{T}(\mathbf{r})$ of Eq. (5),

$$\hat{I}(\mathbf{r}) = -4\Delta = (2i\nabla)^2 = (8m/\hbar^2)\, \hat{T}(\mathbf{r}), \qquad (35)$$

$$\begin{aligned}I[\psi(t)] &= \langle \psi(t)|\hat{I}|\psi(t)\rangle \equiv I(t)\\ &= \int p(\mathbf{r},t)\{[\nabla \ln p(\mathbf{r},t)]^2 + [2\nabla \phi(\mathbf{r},t)]^2\}d\mathbf{r} \equiv \int p(\mathbf{r},t)\, I(\mathbf{r},t)\, d\mathbf{r}\\ &= I[p;t] + 4\int p(\mathbf{r},t)[\nabla \phi(\mathbf{r},t)]^2\, d\mathbf{r} \equiv I[p;t] + I[\phi;t] \equiv I[p,\phi;t]\\ &= I[p;t] + (2m/\hbar)^2 \int p(\mathbf{r},t)^{-1} j(\mathbf{r},t)^2\, d\mathbf{r} \equiv I[p;t] + I[j;t] \equiv I[p,j;t].\end{aligned} \qquad (36)$$

The density-per-electron of resultant gradient information,

$$I(\mathbf{r},t) = [\nabla \ln p(\mathbf{r},t)]^2 + 4[\nabla \phi(\mathbf{r},t)]^2 \equiv I_p(\mathbf{r},t) + I_\phi(\mathbf{r},t), \qquad (37)$$

thus contains the nonclassical (phase) contribution $I_\phi(\mathbf{r}, t)$ proportional to the divergence of the phase current. This generalized measure reflects the (dimensionless) average kinetic energy of electrons $T[\psi] = \langle \psi|\hat{T}|\psi\rangle$:

$I[\psi] = (8m/\hbar^2) T[\psi] \equiv \sigma T[\psi]$.

The phase-continuity relations of Eqs. (29)-(31) involve the nonclassical information-density term $I_\phi(\boldsymbol{r},t)$,

$$\sigma_\phi(\boldsymbol{r},t) \equiv d\phi(\boldsymbol{r},t)/dt = \partial\phi(\boldsymbol{r},t)/\partial t + \nabla \cdot \boldsymbol{J}(\boldsymbol{r},t)$$
$$= \partial\phi(\boldsymbol{r},t)/\partial t + (\hbar/m)[\nabla\phi(\boldsymbol{r},t)]^2 \equiv \partial\phi(\boldsymbol{r},t)/\partial t + \beta I_\phi(\boldsymbol{r},t),$$

$$\beta = \hbar/(4m). \tag{38}$$

This phase-production is seen to involve contributions related to the modulus (R) and phase (ϕ) components of ψ, as well as to the external potential (v):

$$\sigma_\phi(\boldsymbol{r},t) = [\hbar/(2m)]\{R(\boldsymbol{r},t)^{-1}\Delta R(\boldsymbol{r},t) + [\nabla\phi(\boldsymbol{r},t)]^2\} - v(\boldsymbol{r})/\hbar$$
$$\equiv \sigma_\phi[R;\boldsymbol{r},t] + \sigma_\phi[\phi;\boldsymbol{r},t] + \sigma_\phi[v;\boldsymbol{r},t]. \tag{39}$$

Therefore, the quantum gradient-information concept combines the classical contribution $I[p;t]$ and its nonclassical phase/current supplement $I[\phi;t] = I[\boldsymbol{j};t]$. The latter generates a finite information production term

$$\sigma_I(t) = dI(t)/dt = (i/\hbar)\langle \psi(t)|[\hat{H},\hat{I}]|\psi(t)\rangle$$
$$= -(8/\hbar)\int p(\boldsymbol{r},t)\nabla\phi(\boldsymbol{r},t)\cdot\nabla v(\boldsymbol{r})\,d\boldsymbol{r} \equiv \sigma\!\int \boldsymbol{j}(\boldsymbol{r},t)\cdot\boldsymbol{F}(\boldsymbol{r})\,d\boldsymbol{r}$$
$$= dI[\phi;t]/dt = \int[d\phi(\boldsymbol{r},t)/dt]\{\delta I[\phi;t]/\delta\phi(\boldsymbol{r},t)\}\,d\boldsymbol{r}$$
$$= -8\!\int \sigma_\phi(\boldsymbol{r},t)\nabla p(\boldsymbol{r},t)\cdot\nabla\phi(\boldsymbol{r},t)\,d\boldsymbol{r}. \tag{40}$$

This overall source of the resultant gradient-information can be also interpreted in terms of its local continuity equation [see Eq. (27)]:

$$\sigma_I(\boldsymbol{r},t) = dI(\boldsymbol{r},t)/dt = \partial I(\boldsymbol{r},t)/\partial t + \nabla \cdot \boldsymbol{J}_I(\boldsymbol{r},t),$$

$$\boldsymbol{J}_I(\boldsymbol{r},t) = I(\boldsymbol{r},t)\boldsymbol{V}(\boldsymbol{r},t), \quad \nabla \cdot \boldsymbol{J}_I(\boldsymbol{r},t) = \nabla I(\boldsymbol{r},t)\cdot\boldsymbol{V}(\boldsymbol{r},t),$$

where the density-per-electron of information source $\sigma_I(\mathbf{r}, t) = dI[\mathbf{r}(t), t]/dt$ integrates to the resultant production of Eq. (40):

$$\sigma_I(t) = \int p(\mathbf{r}, t)\, \sigma_I(\mathbf{r}, t)\, d\mathbf{r} = \int p(\mathbf{r}, t)\, [\partial I(\mathbf{r}, t)/\partial t + \nabla \cdot \mathbf{J}_I(\mathbf{r}, t)]\, d\mathbf{r}.$$

Therefore, an inclusion of the phase/current component in the resultant gradient-information descriptor generates a nonvanishing source of this (dimensionless) kinetic-energy descriptor. The integral production of the nonclassical information then assumes thermodynamic-like form, of the product of electronic current ("flux") and the external force ("affinity"): $\mathbf{F}(\mathbf{r}) = -\nabla v(\mathbf{r})$.

Similar resultant concepts of the global entropy have also been designed [53-62]. The state resultant *scalar* entropy measure

$$\begin{aligned} S[p, \phi] &= \langle \psi | -\ln p - 2\phi | \psi \rangle = -\int p(\mathbf{r})\, [\ln p(\mathbf{r}) + 2\phi(\mathbf{r})]\, d\mathbf{r} \equiv S[p] + S[\phi] \\ &\equiv \int p(\mathbf{r})[S_p(\mathbf{r}) + S_\phi(\mathbf{r})]\, d\mathbf{r} \equiv \int p(\mathbf{r})\, S(\mathbf{r})\, d\mathbf{r} \end{aligned} \quad (41)$$

combines the classical global entropy of Shannon [Eq. (34)], defined by the density-per-electron $S_p(\mathbf{r}) = -\ln p(\mathbf{r})$,

$$S[p] = -\int p(\mathbf{r})\, \ln p(\mathbf{r})\, d\mathbf{r} \equiv \int p(\mathbf{r})\, S_p(\mathbf{r})\, d\mathbf{r},$$

and its nonclassical, phase complement

$$S[\phi] = -2\int p(\mathbf{r})\, \phi(\mathbf{r})\, d\mathbf{r} = -2\langle \phi \rangle_\psi \equiv \int p(\mathbf{r})\, S_\phi(\mathbf{r})\, d\mathbf{r} \quad (42)$$

exhibiting the density $S_\phi(\mathbf{r}) = -2\phi(\mathbf{r})$. In this overall measure of the state electron "uncertainty" the (positive) classical entropy $S[p]$ in probability distribution is accompanied by the (negative) nonclassical supplement $S[\phi]$ reflecting the state average phase $\langle \phi \rangle_\psi$. The sign of the latter reflects the fact that the presence of a finite electronic current introduces an additional degree of "order" (determinicity), which diminishes the state resultant "disorder" (uncertainty) measure reflected by the overall entropy descriptor.

In the related complex (*vector*) entropy concept, the expectation value of the non-Hermitian operator

$$\hat{S}(r) = -2\ln\psi(r) = -\ln p(r) - 2i\phi(r), \tag{43}$$

these classical and nonclassical contributions constitute the real and imaginary components of such (complex) *vector* measure of the state electronic uncertainty [59]:

$$\vec{S}[\psi] \equiv \langle \psi | \hat{S} | \psi \rangle = S[p] + i S[\phi]. \tag{44}$$

Let us now turn to the generalized gradient measures of electronic uncertainty. Since the nonclassical determinicity contribution $I[\phi] = I[j]$ to the resultant gradient information is nonnegative, its indeterminicity analog

$$M[\phi] = M[j] \equiv -I[\phi] \equiv -I[j]$$

in the resultant gradient entropy [62],

$$M[p, \phi] \equiv I[p] - I[\phi] \equiv M[p] + M[\phi] \equiv I[p] - I[j] \equiv M[p] + M[j], \tag{45}$$

must be nonpositive. Indeed, the nonvanishing current pattern contributes an extra information ("order") contribution into the resultant IT descriptor of electronic state, thus decreasing the overall level of the system gradient entropy ("disorder," uncertainty) contribution.

One further observes that the densities-per-electron of Shannon's entropy, $S_p(r) = -\ln p(r) \geq 0$, and Fisher's information, $I_p(r) = \{\nabla[\ln p(r)]\}^2 \geq 0$, obey the following mutual relations:

$$[\nabla S_p(r)]^2 = \{\nabla[-\ln p(r)]\}^2 = I_p(r) \quad \text{or}$$

$$|\nabla S_p(r)| = \sqrt{I_p(r)} \quad \text{and} \quad S_p(r) = -\ln p(r) = \int_{-\infty}^{r} \sqrt{I_p(u)}\, du. \tag{46}$$

Therefore, by extending these requirements to densities of the resultant entropy and information descriptors one can also design the following geometric measure of the state average overall-entropy content [88]:

$$h[\psi] \equiv \int p(r) h(r)\, dr, \tag{47}$$

defined by the functional density-per-electron $h(r)$ related to the overall gradient-information density $I(r)$ of Eq. (37):

$$[\nabla h(r)]^2 = I(r) \quad \text{or} \quad h(r) = \int_{-\infty}^{r} \sqrt{I(u)}\, du > 0. \tag{48}$$

The emerging geometric measure of the resultant entropy in state ψ then reads:

$$h[\psi] = \int p(r)\{ \int_{-\infty}^{r} \sqrt{I(u)}\, du \}\, dr \equiv \int p(r) h(r)\, dr$$

$$= \int p(r)\{ \int_{-\infty}^{r} \sqrt{\{\nabla[-\ln p(u)]\}^2 + [\nabla \phi(u)]^2}\, du \}\, dr \equiv h[p, \phi]$$

$$= \int [R(r)]^2 \{ \int_{-\infty}^{r} \sqrt{\{\nabla[-2\ln R(u)]\}^2 + [\nabla \phi(u)]^2}\, du \}\, dr \equiv h[R, \phi]. \tag{49}$$

Such a (real) overall measure of the resultant gradient-uncertainty content in the complex quantum state thus satisfies classical relations between densities of the complementary global-entropy and gradient-information descriptors. For the specified electron density or its probability (shape) factor $p(r) = [R(r)]^2$ this functional assumes the minimum value $S[p]$, of the global Shannon entropy in the state probability distribution

$p(r)$, when $\nabla\phi(r, t) = 0$ or $\phi(r, t) = \phi(t)$, i.e., for the stationary quantum states. Therefore, such states can be regarded as the geometric phase-equilibria, which minimize $h[\psi]$ for the state "frozen" modulus or probability distributions.

A final comment concerns the relation between the density-per-electron $I(r)$ of overall gradient information and the logarithmic derivative $L[\psi] \equiv \nabla\ln\psi$ of the system wavefunction $\psi(r)$ [see Eq. (20)]:

$$L[\psi(r)] = \psi(r)^{-1}\nabla\psi(r) = \nabla\ln R(r) + i\nabla\phi(r)$$
$$= \tfrac{1}{2}\nabla\ln p(r) + i\nabla\phi(r).$$

This complex gradient separates the classical (real) part due to the probability density and the nonclassical (imaginary) term generated by the gradient of the state phase. The gradient-information density $I(r)$ thus reflects the derivative magnitude. Indeed, a reference to Eq. (37) directly shows that

$$I(r) = [\nabla\ln p(r)]^2 + 4[\nabla\phi(r)]^2 = I_p(r) + I_\phi(r) = 4|L[\psi(r)]|^2.$$

The classical $[I_p(r)]$ and nonclassical $[I_\phi(r)]$ contributions to the information density $I(r)$, or the related kinetic-energy densities, thus reflect the squared modulus of the local logarithmic derivative $L[\psi(r)]$ of the system electronic wavefunction.

5. OPEN SYSTEMS AND BONDED MOLECULAR FRAGMENTS

Consider the pure (complex) electronic wavefunctions of N_R electrons (integer) describing the externally-closed, equilibrium reactive system $R = (A^* \mid B^*) \equiv R^*$ composed of the mutually-open reactants $\alpha^* \in (A^*, B^*)$ (see Figure 2b),

$$\Psi(N_R) = M(N_R) \exp[i\Phi(N_R)], \tag{50}$$

where the real functions $M(N_R)$ and $\Phi(N_R)$ stand for its modulus and phase components, respectively. We shall also examine the reference states of polarized subsystems $\alpha^+ \in (A^+, B^+)$, the mutually- and externally-closed, substrates in $R^+ = (A^+|B^+)$,

$$\{\psi_j^i(\alpha) = R_j^i(\alpha) \exp[i\phi_j^i(\alpha)] \equiv \psi_j(N_\alpha^i) \equiv R_\alpha^j(N_\alpha^i) \exp[i\phi_\alpha^j(N_\alpha^i)]\}. \tag{51}$$

It follows from DFT [40, 41] that the ground-state energetics of the molecular reactive system R is uniquely determined by the system density $\rho_R(r)$:

$$E[N_R, v] = E_v[\rho_R] = \int \rho_R(r) v(r) dr + F[\rho_R]. \tag{52}$$

The externally-open fragment $R^*(\mu_R)$ of the (macroscopic) composite system $[\mathcal{R}(\mu_R)|R^*(\mu_R)]$, in contact with electron reservoir $\mathcal{R}(\mu_R)$ [see Eq. (2)], is described by the *mixed* quantum state, a statistical mixture of the pure R-states $\{\Psi_l^k \equiv \Psi_l(N^k)\}$ for different (integer) numbers $\{N^k\}$ of electrons in R,

$$\{\Psi_l^k = M_l(N^k) \exp[i\Phi_l(N^k)] \equiv M_l^k \exp(i\Phi_l^k)\}, \tag{53}$$

generating the associated state densities and currents:

$$\{\rho_l^k(r) = \rho[\Psi_l^k; r] = \rho_l(N^k; r) \equiv N^k p_l^k(r)\},$$

$$\{j_l^k(r) = j[\Psi_l^k; r] = j_l(N^k; r) = (\hbar/m) p_l^k(r) \nabla \Phi_l^k(r)\}. \tag{54}$$

The ensemble average quantities are then defined by the "molecular" density operator

$$\sum_k \sum_l |\Psi_l^k\rangle P_l^k \langle \Psi_l^k|, \quad \sum_k (\sum_l P_l^k) \equiv \sum_k P^k = 1, \tag{55}$$

where state-probabilities $\mathbf{P}(\mu_R, T) \equiv \{P_l^k(\mu_R, T) \equiv P_l^k\}$ reflect the system thermodynamic conditions (μ_R, T): chemical potential μ_R of the (external) electron reservoir and temperature T of the "heat bath". The average (fractional) number of electrons

$$N_R^* = \int \rho_R^* d\mathbf{r} = \sum_k P^k N^k, \qquad \rho_R^*(\mathbf{r}) = \sum_k \sum_l P_l^k \rho_l^k(\mathbf{r}), \qquad (56)$$

ultimately reflects the external CT between R and its reservoir in $(\mathscr{R}^*|R^*)$. In such an equilibrium thermodynamic state of the externally-open reactive system the average electron density $\rho_R^*(\mathbf{r})$ determines the resultant electronic structure of "being", while the associated ensemble-average current pattern

$$\mathbf{j}_R^*(\mathbf{r}) = \sum_k \sum_l P_l^k \mathbf{j}_l^k(\mathbf{r}) = (\hbar/m) \sum_k \sum_l P_l^k p_l^k(\mathbf{r}) \nabla \Phi_l^k(\mathbf{r}) \qquad (57)$$

describes the resultant structure of "becoming".

Due to the (fractional) CT between the mutually-open subsystems in the externally-closed $R = (A^*|B^*)$, their equilibrium densities $\{\rho_\alpha^*(\mathbf{r}) = N_\alpha^* p_R(\mathbf{r})\}$ conserving the integer value of $N_R = N_A^* + N_A^*$, pieces of the "molecular" density

$$\rho_R(\mathbf{r}) = \sum_\alpha \rho_\alpha^*(\mathbf{r}) = (\sum_\alpha N_\alpha^*) p_R(\mathbf{r}) = N_R p_R(\mathbf{r}), \qquad (58)$$

also exhibit fractional numbers of electrons $\{N_\alpha^* = \int \rho_\alpha^* d\mathbf{r}\}$ and require an ensemble description. Such open fragments reflect "molecular" probability distribution $p_R(\mathbf{r})$, the *shape*-function of $\rho_R(\mathbf{r}) = N_R p_R(\mathbf{r})$.

The mixed state of subsystem α^* is defined by the fragment density operator

$$\sum_i \sum_j |\psi_j^i(\alpha)\rangle P_j^i(\alpha) \langle \psi_j^i(\alpha)|, \qquad \sum_i [\sum_j P_j^i(\alpha)] \equiv \sum_i P^i(\alpha) = 1, \qquad (59)$$

where the equilibrium probabilities in α^*, $\{P_j^i(\alpha) \equiv P_j^i(\mu_\alpha, T_\alpha)\} \equiv \mathbf{P}(\alpha)$, are again shaped by the subsystem (external) thermodynamic conditions $\{(\mu_\alpha,$

$T_\alpha)\}$. It defines the statistical mixture of the fragment pure-states $\{|\psi_j^i(\alpha)\rangle = |\psi_j(N_\alpha^i)\rangle\}$, for the integer numbers of electrons $\{N_\alpha^i\}$ in α^+, generating the subsystem electron densities $\{\rho_j^i(\alpha;r) \equiv N_\alpha^i p_j^i(\alpha;r)\}$ and currents

$$\{j_j^i(\alpha;r) = j[\psi_j^i(\alpha); r] = j_j(N_\alpha^i; r) = (\hbar/m) p_j^i(\alpha;r) \nabla \phi_j^i(\alpha;r)\}. \tag{60}$$

Thermodynamic equilibrium state corresponding to the density operator of Eq. (59) gives rise to the following average descriptors of the open fragment α^*: its electron density,

$$\rho_\alpha^*(r) = \Sigma_i \Sigma_j P_j^i(\alpha) \rho_j^i(\alpha;r) = N_\alpha^* p_R(r), \tag{61}$$

overall number of electrons,

$$N_\alpha^* = \int \rho_\alpha^*(r) dr = \Sigma_i [\Sigma_j P_j^i(\alpha)] N_\alpha^i \equiv \Sigma_i P_\alpha^i N_\alpha^i, \tag{62}$$

and the associated ensemble-mean distributions of the average probability and current:

$$p_\alpha^*(r) = \Sigma_i \Sigma_j P_j^i(\alpha) p_j^i(\alpha;r),$$

$$j_\alpha^*(r) = \Sigma_i \Sigma_j P_j^i(\alpha) j_j^i(\alpha;r) = (\hbar/m) \Sigma_i \Sigma_j P_j^i(\alpha) p_j^i(\alpha;r) \nabla \phi_j^i(\alpha;r). \tag{63}$$

The open fragment α^* is thus described by the mixed quantum state in the subsystem grand ensemble, corresponding to its average phase

$$\phi_\alpha^*(r) = \Sigma_i \Sigma_j P_j^i(\alpha) \phi_j^i(\alpha;r). \tag{64}$$

The fractional number of electrons N_α^* in α^* then reflects the inter-fragment CT and the "molecular" shape-factor $p_R(r)$ of the electronic density in R^* [Eq. (58)].

One further observes that the two basis sets,

$$|\Psi\rangle = \{|\Psi_l^k(R)\rangle\} \quad \text{and}$$

$$|\chi(A|B)\rangle = (|\psi(\alpha)\rangle \otimes |\psi(\beta)\rangle = \{|\psi_j^i(\alpha)\psi_{j'}^{i'}(\beta)\rangle \equiv |\chi_{j,j'}^{i,i'}(\alpha|\beta)\rangle\}), \quad (65)$$

constitute alternative (complete) "molecular" sets,

$$|\Psi\rangle\langle\Psi| = |\chi(A|B)\rangle\langle\chi(A|B)| = \Sigma_\alpha |\psi(\alpha)\rangle\langle\psi(\alpha)| = 1, \quad (66)$$

of orthonormal quantum states of N_R electrons:

$$\langle\Psi|\Psi\rangle = \{\delta_{k,k'}\delta_{l,l'}\} = \mathbf{I} \quad \text{and}$$

$$\langle\psi(\alpha)|\psi(\beta)\rangle = \delta_{\alpha,\beta}\{\delta_{i,i'}\delta_{j,j'}\} = \delta_{\alpha,\beta}\mathbf{I}(\alpha). \quad (67)$$

Here,

$$\begin{aligned}\langle\chi(A|B)|\chi(A|B)\rangle &= \{\langle\chi_{j,j'}^{i,i'}(\alpha|\beta)|\chi_{n,n'}^{m,m'}(\alpha'|\beta')\rangle \\ &= \langle\psi_j^i(\alpha)\psi_{j'}^{i'}(\beta)|\psi_n^m(\alpha')\psi_{n'}^{m'}(\beta')\rangle\} \\ &= \{\langle\psi_j^i(\alpha)|\psi_n^m(\alpha')\rangle\langle\psi_{j'}^{i'}(\beta)|\psi_{n'}^{m'}(\beta')\rangle \\ &= (\delta_{\alpha,\alpha'}\delta_{i,m}\delta_{j,n})(\delta_{\beta,\beta'}\delta_{i',m'}\delta_{j',n'})\} \\ &= \{[\delta_{\alpha,\alpha'}\mathbf{I}(\alpha)][\delta_{\beta,\beta'}\mathbf{I}(\beta)]\} = \mathbf{I}(A|B),\end{aligned}$$

where $\mathbf{I}$, $\mathbf{I}(A|B)$ and $\mathbf{I}(\alpha)$ stand for unit matrices determining the corresponding identity transformations. These two basis sets are thus related by the unitary transformation $\mathbf{U} = \langle\chi|\Psi\rangle$:

$$|\Psi\rangle = |\chi\rangle\mathbf{U} \quad \text{or} \quad |\Psi\rangle = |\chi\rangle\mathbf{U}^\dagger, \quad \mathbf{U}^\dagger\mathbf{U} = \mathbf{U}\mathbf{U}^\dagger = \mathbf{I}. \quad (68)$$

Therefore, the matrix representations of quantum observables in these two bases are related by similarity transformation, which does not affect traces defining the ensemble-averages of physical properties. It indeed follows from Eq. (18) that thermodynamic equilibria in

$$[\mathcal{R}_A^*(\mu_R)|A^*(\mu_R)|B^*(\mu_R)|\mathcal{R}_B^*(\mu_R)] \quad \text{and} \quad [\mathcal{R}(\mu_R)|R^*(\mu_R)]$$

are identical (see Figure 2b), since they both correspond to the same set of state-parameters in the molecular grand ensemble: $\mu = \mu_R$, T and v.

6. ADDITIVE ENTROPY/INFORMATION DESCRIPTORS OF SUBSYSTEMS

The reactant densities $\{\rho_\alpha(r)\}$, resulting from the given partition of the "molecular" electron density $\rho_R(r)$ in R as a whole, combine additively:

$$\rho_R(r) \equiv \rho_A(r) + \rho_B(r) = \sum_\alpha \rho_\alpha(r), \qquad (69)$$

and so one expects of the "proper" IT descriptors in thermodynamic-like treatment of reactive systems. Let us first consider the resultant entropy functional $S[p, \phi]$ [Eq. (41)], formulated in terms of the (unity-normalized) probability density $p_R(r) = \rho_R(r)/N_R$, $\int p_R(r)\,dr = 1$, i.e., the electron density per an electron.

In *many*-electron systems the relevant additive extension of the overall entropy follows naturally from multiplying $S[p, \phi]$ by the overall number of electrons $N_R = \int \rho_R(r)\,dr$ and recognizing the overall measure of resultant entropy in $\rho_R(r)$ as the *average* (integral) descriptor resulting from the spatial averaging of the *entropy-per-electron* density

$$S_R(r) = -\ln p_R(r) + 2\phi_R(r), \qquad (70)$$

with the local "weight" of the electron density in r:

$$\begin{aligned}
S[N_R, p_R, \phi_R] = N_R\, S[p_R, \phi_R] &\equiv S[\rho_R, \phi_R] = \int \rho_R(r)\, S_R(r)\, dr \\
&= -\int \rho_R(r)\, [\ln p_R(r) + 2\phi_R(r)]\, dr \\
&\equiv \sum_\alpha \{S[\rho_\alpha, p_R] + S[\rho_\alpha, \phi_R]\}.
\end{aligned} \qquad (71)$$

Therefore, the density partition of Eq. (69) implies the associated (additive) division of the overall entropy functional:

$$S[N_R, p_R, \phi_R] \equiv S[\rho_R, \phi_R] = \sum_\alpha \int \rho_\alpha(r) S_R(r) \, dr \equiv \sum_\alpha S[\rho_\alpha, p_R, \phi_R]. \qquad (72)$$

The molecular (*many*-electron) complex-entropy descriptor of Eq. (44) partitions similarly:

$$\begin{aligned}N_R (S[p_R] + i S[\phi_R]) &= -\int \rho_R(r) \, [\ln p_R(r) + 2i\phi_R(r)] \, dr \\ &\equiv S[\rho_R, p_R] + i \, S[\rho_R, \phi_R] \\ &= -\sum_\alpha \int \rho_\alpha(r)[\ln p_R(r) + 2i\phi_R(r)] \, dr \\ &\equiv \sum_\alpha \{S[\rho_\alpha, p_R] + i \, S[\rho_\alpha, \phi_R]\}. \end{aligned} \qquad (73)$$

The geometric entropy of Eq. (49) gives rise to the following additive contributions due to *many*-electron subsystems:

$$\begin{aligned}h[N_R, p_R, \phi_R] &= N_R \, h[p_R, \phi_R] = \int \rho_R(r) \, h_R(r) \, dr \equiv h[\rho_R, \phi_R] \\ &= \sum_\alpha \int \rho_\alpha(r) \, h_R(r) \, dr \equiv \sum_\alpha h[\rho_\alpha, p_R, \phi_R]. \end{aligned} \qquad (74)$$

One similarly generalizes the gradient descriptors of resultant information/entropy content [Eqs. (36) and (45)]. One interprets the overall functional as resulting from the (electron-density weighted) densities-per-electron of resultant IT measures, functions of the probability density and phase in the whole reactive system R:

$$\begin{aligned} I[N_R, p_R, \phi_R] &= N_R \, I[p_R, \phi_R] \equiv I[\rho_R, \phi_R] \\ &= \sum_\alpha \int \rho_\alpha(r) \, I_R(r) \, dr \equiv \sum_\alpha I[\rho_\alpha, p_R, \phi_R], \\ I_R(r) &= [\nabla \ln p_R(r)]^2 + 4 \nabla \phi_R(r)]^2; \end{aligned} \qquad (75)$$

$$\begin{aligned} M[N_R, p_R, \phi_R] &= N_R \, M[p_R, \phi_R] \equiv M[\rho_R, \phi_R] \\ &= \sum_\alpha \int \rho_\alpha(r) \, M_R(r) \, dr \equiv \sum_\alpha M[\rho_\alpha, p_R, \phi_R], \\ M_R(r) &= [\nabla \ln p_R(r)]^2 - 4 \nabla \phi_R(r)]^2. \end{aligned} \qquad (76)$$

To summarize, the electron density factor replaces the probability weight in *many*-electron generalizations of the average (additive) IT descriptors of electronic states in molecular systems and their fragments. The probability and phase "intensities" defining the entropy/information density-per-electron can be either molecular or reactant in character, when they determine the average descriptors of the whole reactive system and of its substrate fragments, respectively.

7. PHASE EQUALIZATION IN BONDED FRAGMENTS

For simplicity, let us first assume *one*-electron subsystems of the acceptor-donor reactive system R = A—B, $\{N^i(\alpha^+) = N_\alpha^0 = 1\}$, when $\{\rho_\alpha = p_\alpha\}$ and $N_R = 2$, in their lowest pure (ground) states $\{\psi_0^i(\alpha) = \Psi_\alpha^+(i)\}$, describing the mutually-*closed* (*c*) (disentangled), reactants in the external potential v due to the nuclei in both subsystems (see Figure 1b),

$$R_c^+ \equiv [A^+(\rho_A^+[v]) | B^+(\rho_B^+[v])], \tag{77}$$

generating the associated (polarized) electron densities

$$\{\rho_0^i(\alpha) \equiv \rho_\alpha^+[N_\alpha^0, v] = (R_\alpha^+)^2\}.$$

The assumed *one*-electron character of both reactants also implies the equal condensed probabilities of these molecular fragments:

$$\{P_\alpha^+ = N_\alpha^0/N_R = 1/2\}.$$

This "microcanonical" scenario of the externally- and mutually-*closed* subsystems thus corresponds to a special case of the ensemble description: $P_{j=0}^i(\alpha) = 1$, $\{P_{j\neq 0}^i(\alpha) = 0\}$. The *two*-electron wavefunction of R_c^+ is then given by the product of subsystem states $\{\Psi_\alpha^+(i) = R_\alpha^+(i) \exp[i\phi_\alpha^+(i)]\}$,

$$\Psi_c^+(1,2) = \Psi_A^+(1)\,\Psi_B^+(2) \equiv M_R^c(1,2) \exp[i\,\Phi_R^c(1,2)], \tag{78}$$

marking their distinguishable groups of electrons. The *two*-electron parts of $\Psi_c^+(1, 2)$, the wavefunction modulus, $M_R^c(1, 2)$, and phase, $\Phi_R^c(1, 2)$, are thus given by the product and sum of the corresponding subsystem components:

$$M_R^c(1, 2) = R_A^+(1) R_B^+(2) \quad \text{and} \quad \Phi_R^c(1, 2) = \phi_A^+(1) + \phi_B^+(2).$$

The following question now arises: what is the overall phase $\Phi_o^+(N_R)$ which characterizes the mutually-open (*o*) (entangled) subsystems in the *open* polarized system

$$R_o^+ = (A^+[\rho_A^+] \vdots B^+[\rho_B^+]) \tag{79}$$

exhibiting the same overall ("molecular") density

$$\rho_R^+ \equiv \rho_A^+ + \rho_B^+ \equiv 2p_R^+,$$

resulting from contributions $\{\rho_\alpha^+\}$ due to the "frozen" polarized reactants? One observes that the system then contains indistinguishable electrons described by the antisymmetrized product of Eq. (78), the associated Slater determinant:

$$\Psi_o^+(1, 2) = |\Psi_A^+ \Psi_B^+|. \tag{80}$$

As an illustration, let us recall the plane-wave type equidensity orbitals (EO) $\psi_k[p] = \{\psi_l[p]\}$ for the given probability distribution $p(r)$,

$$\psi_k[p; r] = \{\psi_l[p; r] = p(r)^{1/2} \exp(i k_l \cdot f[p; r])\}, \tag{81}$$

involving density-dependent (local) *phase*-vector $f[p_R; r]$ safeguarding EO orthogonality. They are used in the Harriman-Zumbach-Maschke (HZM) construction of DFT [89-92], of Slater determinants yielding the prescribed electron density $\rho(r) = Np(r)$:

$$\Psi_k[p] = |\psi_1[p]\, \psi_2[p]\, \ldots\, \psi_N[p]| \equiv \det(\psi_k[p]). \tag{82}$$

The EO states associated with the resultant electron distribution of R_c^+ exhibit the effective modulus function

$$R_o^+(r) = [p_R^+(r)]^{1/2} = \{[\rho_A^+(r) + \rho_B^+(r)]/2\}^{1/2} \equiv [\rho_a^+(r)]^{1/2}, \tag{83}$$

determined by the *arithmetic* average $\rho_a^+(r)$ of the polarized subsystem densities.

The same effective *one*-electron modulus results from the *two*-electron distribution $\rho_R(r_1, r_2) = 2\, M_R^c(r_1, r_2)^2$. Indeed, calculating the probability of finding any of the two electrons at the specified location r gives the square of $R_o^+(r)$:

$$p_R^+(r) = P_A^+ \rho_B^+(r) + P_B^+ \rho_A^+(r) = \rho_a^+(r) = R_o^+(r)^2. \tag{84}$$

In what follows we address the problem of a phase "signature" of "entangling" (mutually-opening) the reactant subsystems, without affecting their electron densities, in modulus-frozen "molecular" state Ψ_o^+ of R_o^+,

$$\Psi_o^+ \equiv \Psi_o[\{\rho_a^+\}] = M_o^+(N_R)\, \exp[i\Phi_o^+(N_R)],$$

$$M_o^+(N_R) = R_A^+(N_A^0)\, R_B^+(N_B^0). \tag{85}$$

The specific HZM construction of such complex states yielding the specified electron density has been discussed elsewhere [89-92]. We also examine a consistency of the equilibrium implications resulting from alternative measures of the overall content of the state entropy or information.

One first observes, that the mutual opening of two reactants, for the frozen overall density $\rho_R^+ = N_R\, p_R^+$, affects both the effective densities and phases of wavefunctions describing the bonded fragments in R_o^+:

$$\rho_a^+[R_o^+] \to \rho_a^*[N_R \rho_R^+] \equiv \rho_a^o = N_a^o\, p_a^o,$$

$$N_\alpha^* = \int \rho_\alpha^o d\mathbf{r} = \int \rho_\alpha^+ d\mathbf{r} \equiv N_\alpha^+, \quad \text{and}$$

$$\phi_\alpha^+[R_o^+] \to \phi_\alpha^*[\rho_R^+] \equiv \phi_\alpha^o. \tag{86}$$

In R_o^+ each fragment density ρ_α^o extends over the probability distribution of the whole ("molecular") reactive system [see Eq. (13)]. This piece of ρ_R^+ just constitutes the (fractional) N_α^o-multiple of the "molecular" probability distribution, in the reactive system as a whole, the square-root of which defines the modulus factor of the associated EO [see Eq. (81)]. Therefore, this orbital representation provides a compact description of CT phenomena between the mutually-open subsystems: displacements in the fragment electron populations manipulate only the occupations of "molecular" EO. One further recognizes the additivity of alternative entropy/information functionals for subsystems in combining the fragment descriptors into resultant measures for the whole R_o^+. Moreover, since no hypothetical boundary separates the two fragments in this (internally) open system corresponding to the "frozen" density $\rho_R^+ = N_R p_R^+$ of the whole R^+, one recalls that each (open) subsystem in this "entangled" reactive system, $\{\alpha^*[R_o^+] \equiv \alpha_o^+\}$, exhausts the "molecular" probability density p_R^+ of Eq. (13):

$$p_\alpha^*[R_o^+] = p_R^+ \equiv p_\alpha^o \quad \text{or} \quad \rho_\alpha^*[R_o^+] = N_\alpha^* p_R^+ = N_\alpha^* p_\alpha^* \equiv \rho_\alpha^o. \tag{87}$$

Consider first the alternative resultant-entropy functionals. The scalar entropy of Eqs. (41), (71) and (72) gives the following combination rule determining the resultant entropy in terms of the additive fragment descriptors:

$$S[\rho_R^+, \Phi_o^+] = \sum_\alpha S[\rho_\alpha^o, \phi_\alpha^o] \quad \text{or}$$

$$\sum_\alpha \int \rho_\alpha^o \{2(\Phi_o^+ - \phi_\alpha^o) + \ln(p_R^+/p_\alpha^o)\} d\mathbf{r} = 2\sum_\alpha \int \rho_\alpha^o (\Phi_o^+ - \phi_\alpha^o) d\mathbf{r} = 0, \tag{88}$$

where we have taken into account Eq. (87). This global measure of the state entropy thus predicts equalization of phases of two mutually-open (bonded) reactants at the global phase Φ_o^+ describing the whole system of the two entangled subsystems:

$$\phi_\alpha^o[\rho_R^+] = \Phi_o^+[\rho_R^+], \quad \alpha = A, B. \tag{89}$$

This fragment phase-equalization rule, reminiscent of the chemical potential equalization principle of Eq. (14), confirms the "intensive" character of the state phase "variable."

The same substrate phase-equalization resulting from the mutual-opening of bonded subsystems follows from the remaining, alternative descriptors of the state overall uncertainty (entropy). For example, the complex entropy of Eq. (44) directly gives:

$$\sum_\alpha \int \rho_\alpha^o \{2i(\Phi_o^+ - \phi_\alpha^o) + \ln(p_R^+/p_\alpha^o)\} d\mathbf{r} = 2i\sum_\alpha \int \rho_\alpha^o (\Phi_o^+ - \phi_\alpha^o) \, d\mathbf{r} = 0$$

or

$$\phi_\alpha^o = \Phi_o^+, \quad \alpha = A, B. \tag{90}$$

The consistency equations following from the *geometric*-entropy (+) measure of Eq. (49) and the *gradient*-entropy (−) descriptor of Eq. (45) similarly predict

$$\sum_\alpha \int \rho_\alpha^o \{\pm 4\nabla(\Phi_o^+ - \phi_\alpha^o) \cdot \nabla(\Phi_o^+ + \phi_\alpha^o) + \nabla \ln(p_R^+/p_\alpha^o) \cdot \nabla \ln(p_R^+ p_\alpha^o)\} d\mathbf{r}$$
$$= 4\sum_\alpha \int \rho_\alpha^o \nabla(\Phi_o^+ - \phi_\alpha^o) \cdot \nabla(\Phi_o^+ + \phi_\alpha^o) \, d\mathbf{r} = 0$$

or

$$\phi_\alpha^o = \Phi_o^+ + const. \equiv \Phi_o^+, \quad \alpha = A, B, \tag{91}$$

since a constant phase component is immaterial in QM and can be set to zero.

Consider finally the functional of Eq. (36), for the resultant gradient-information. The information additivity [Eq. (75)] again requires

$$I[\rho_R^+, \Phi_o^+] = \sum_\alpha I[\rho_\alpha^o, p_\alpha^o, \phi_\alpha^o] \quad \text{or}$$

$$\sum_\alpha \int \rho_\alpha^o \{4\nabla(\Phi_o^+ - \phi_\alpha^o) \cdot \nabla(\Phi_o^+ + \phi_\alpha^o) + \nabla \ln(p_R^+/p_\alpha^o) \cdot \nabla \ln(p_R^+ p_\alpha^o)\} dr$$
$$= 4\sum_\alpha \int \rho_\alpha^o \nabla(\Phi_o^+ - \phi_\alpha^o) \cdot \nabla(\Phi_o^+ + \phi_\alpha^o) \, dr = 0, \quad (92)$$

where we have again recognized Eq. (87). This condition again implies the substrate phase-equalization of Eq. (91).

Therefore, one concludes that a mutual opening (entanglement, bonding) of reactant subsystems, when the fragment electron density exhausts the "molecular" probability distribution, indeed implies simultaneous equalizations of the fragment *modulus*- and *phase*-intensities at levels describing the reactive system as a whole. The chemical potentials of open subsystems then reflect the molecular electronegativity of a common electron reservoir in the equilibrium composite system $\mathcal{M}_R^*$ (see Figure 2b), while phases of their polarized states reflect the common, "molecular" phase describing the whole reactive system R_o^+.

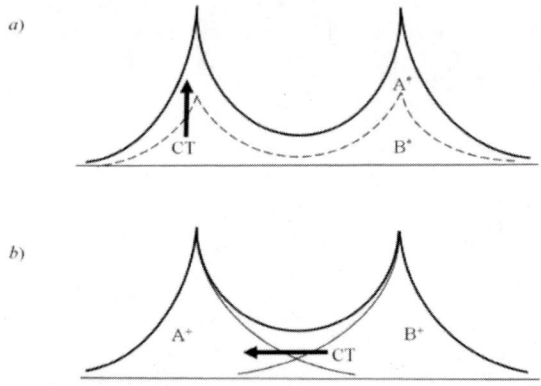

Figure 3. The "vertical" (Panel *a*) and "horizontal" (Panel *b*) B→A CT in the donor–acceptor systems involving the electron-occupation displacements of EO representing the delocalized distributions of bonded (mutually-open) equilibrium fragments $\{\alpha^*\}$ in $R^* = (A^* \vdots B^*)$ and the localized densities of nonbonded (mutually-closed) polarized reactants $\{\alpha^+\}$ in $R^+ = (A^+ | B^+)$, respectively.

As schematically illustrated in Figure 3, the CT between the (mutually) *open* subsystems has the "vertical" character, of the orbital-occupation displacement between the delocalized EO defined by the same (molecular) probability distribution. Accordingly, a transfer of electrons between the (polarized) *closed* subsystems is of the "horizontal" character, of a shift in occupations of the localized EO on molecular fragments, defined by the reactant probability densities.

It thus follows from the above phase-equalization principle that the equilibrium states in all mutually-open fragments are characterized by a common phase descriptor. Therefore, in the "frozen" promolecular reference

$$R^0 = ([a_1^0 | a_2^0 | \ldots] | [b_1^0 | b_2^0 | \ldots])$$

the state of each (mutually-closed) constituent unit of both reactants exhibits its own phase descriptor,

$$\phi_{a1}^0 \neq \phi_{a2}^0 \neq \ldots \neq \phi_{b1}^0 \neq \phi_{b2}^0. \tag{93}$$

In the polarized system

$$R^+ = (a_1^+ | a_2^+ | \ldots | b_1^+ | b_2^+ | \ldots)$$

phases of the equilibrium states describing constituent parts of each reactant are equalized,

$$\phi_{a1}^+ = \phi_{a2}^+ = \ldots \equiv \Phi_A^+ \neq \phi_{b1}^+ = \phi_{b2}^+ = \ldots \equiv \Phi_B^+, \tag{94}$$

while in the fully relaxed, global-equilibrium complex

$$R^* = (a_1^* | a_2^* | \ldots | b_1^* | b_2^* | \ldots)$$

the phase equalization extends over the whole reactive system:

$$\phi_{a1}^* = \phi_{a2}^* = \ldots \equiv \Phi_A^* = \phi_{b1}^* = \phi_{b2}^* = \ldots \equiv \Phi_B^* = \Phi_R. \tag{95}$$

8. EQUIDENSITY-ORBITAL SYSTEMS

Consider again the HZM [89-92] determinant of Eq. (82), constructed using the occupied EO of Eq. (81),

$$\psi_k[p;r] = \{\psi_l[p;r] = p(r)^{1/2} \exp\{i k_l \cdot f[p;r]\}\},$$

which reconstruct the molecular probability density $p(r) = \rho(r)/N$, or the equilibrium EO, phase-transformed analog:

$$\psi_k^{eq.}[p;r] = \{\psi_l^{eq.}[p;r] = p(r)^{1/2} \exp\{i(k_l \cdot f[p;r] + \phi[p;r])\} \\ = p(r)^{1/2} \exp\{i[F_l(r) + \phi(r)]\} \equiv p(r)^{1/2} \exp\{i\Phi_l(r)\}. \tag{96}$$

The latter involves the local "thermodynamic" phase

$$\phi[p;r] = -\langle k[p] \rangle \cdot f[p;r] \equiv \phi(r), \quad \langle k[p] \rangle = N^{-1} \sum_l k_l[p], \tag{97}$$

common in all occupied EO, determined from the supplementary principle of the state minimum overall information [91, 92]. It exhibits the resultant phase $\Phi_l(r)$, combining the EO "orthogonality" phase $F_l(r) = k_l \cdot f[p;r]$ and "thermodynamic" phase supplement $\phi(r)$:

$$\Phi_l(r) = (k_l - \langle k \rangle) \cdot f[p;r] \equiv \delta k_l \cdot f[p;r]. \tag{98}$$

It is shaped by the orbital wave-vector displacement $\{\delta k_l = k_l - \langle k \rangle\}$. This equilibrium phase also results from optimizing the weighted average of EO wave-numbers, $\langle k[P] \rangle = \sum_l P_l k_l[p]$, $\sum_l P_l = 1$, for the configuration fixed overall wave-vector $K[p] = \sum_l k_l[p]$. Indeed, the optimum solutions of variational principle for the unknown probability weights $P = \{P_l\}$,

$$\delta\{\langle k[P]\rangle - \lambda K[p]\} = 0,$$

recovers the average wave-vector of Eq. (97): $\{P_l = \lambda = N^{-1}\}$.

As indicated in Eq. (97), the construction vector function $f[p; r] \equiv f(r) = \{f_x(x, y, z), f_y(y, z), f_z(z)\}$ is uniquely specified by the state probability density $p(r)$, $f(r) = f[p; r]$, with the Jacobian of the HZM transformation $r \to f(r)$, defined by product of diagonal coordinate derivatives $\{\partial f_\xi(r)/\partial x_\xi\}$,

$$\partial f(r)/\partial r = [\partial f_x(x, y, z)/\partial x] [\partial f_y(y, z)/\partial y] [\partial f_z(z)/\partial z],$$

related to conditional probabilities of the spatial coordinates of an electron [89-92].

The EO contributions $\{j_l[p; r]\}$ to the system overall electronic current $j_k[p; r] = \sum_l j_l[p; r]$,

$$j_l[p; r] = (\hbar/m)\, p(r)\, \nabla F_l(r) = (\hbar/m)\, p(r)\, k_l\, \nabla \cdot f[p; r], \tag{99}$$

involve the divergence of HZM vector function $f[p; r]$ related to the probability of the *union* of electronic coordinates, $P[U(r)] = P(x \vee y \vee z)$,

$$\nabla \cdot f[p; r] = 2\pi\, P[U(r)],$$

represented by the sum (envelope) of probability components shown in Figure 4:

$$\begin{aligned}
P[U(r)] &= p(x|y, z) + p(y|z) + p(z) \\
&= p(x, y, z)/p(y, z) + p(y, z)/p(z) + p(z) \\
&= p(r)/\int p(r)\, dx + \int p(r)\, dx/\iint p(r)\, dx\, dy + \iint p(r)\, dx\, dy.
\end{aligned} \tag{100}$$

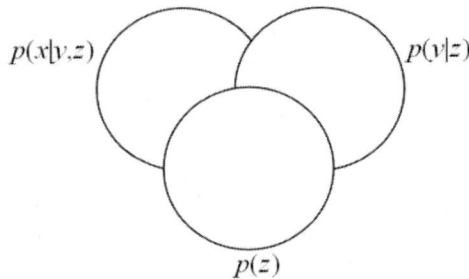

Figure 4. Partition of the probability density $P[U(r)] \equiv P(x \vee y \vee z)$ for the *union* $U(r)$ of the spatial coordinates $\{x_\alpha\}$ of an electron corresponding to the envelope of three dependent (overlapping) probability circles $\{p(x_\alpha)\}$ for separate Cartesian coordinates of the electron position vector $\mathbf{r} = (x \wedge y \wedge z) \equiv (x, y, z)$, the *product* of coordinate events, into partial areas representing $p(z)$ and supplementary conditional probabilities $p(y|z)$ and $p(x| y \wedge z) \equiv p(x| y, z)$.

One recalls that in HZM construction the components $\{f_\xi\}$ of the vector function $\mathbf{f} = \mathbf{i}f_x + \mathbf{j}f_y + \mathbf{k}f_z$ determining the EO orthogonality phase $F_l(r)$ are generated as (indefinite) integrals of the electron probability function $p(\mathbf{r}) = p(x \wedge y \wedge z) \equiv p(x, y, z)$ and its partial (definite) integrals $p(y, z) = \int p(x, y, z)\, dx$ and $p(z) = \int p(y, z)\, dy$:

$$f_x(x,y,z) = 2\pi\, p(y,z)^{-1} \int_{-\infty}^{x} dx'\, p(x',y,z) ,$$

$$f_y(y,z) = 2\pi\, p(z)^{-1} \int_{-\infty}^{y} dy'\, p(y',z), \qquad f_z(z) = 2\pi \int_{-\infty}^{z} dz'\, p(z') . \qquad (101)$$

They determine the Jacobian determinant for HZM transformation $\mathbf{r} \to \mathbf{f}(r)$,

$$\frac{\partial \mathbf{f}}{\partial \mathbf{r}} = \begin{vmatrix} \frac{\partial f_x}{\partial x} & 0 & 0 \\ \frac{\partial f_x}{\partial y} & \frac{\partial f_y}{\partial y} & 0 \\ \frac{\partial f_x}{\partial z} & \frac{\partial f_y}{\partial z} & \frac{\partial f_z}{\partial z} \end{vmatrix} = \left(\frac{\partial f_x}{\partial x}\right)\left(\frac{\partial f_y}{\partial y}\right)\left(\frac{\partial f_z}{\partial z}\right), \qquad (102)$$

by using the diagonal coordinate derivatives of components (f_x, f_y, f_z) expressed in terms of conditional probabilities $p(x|y,z)$ and $p(y|z)$, or $p(z)$:

$$\left(\frac{\partial f_x}{\partial x}\right) \equiv 2\pi\, p(x|y,z) = 2\pi p(x,y,z)/p(y,z) = 2\pi\, p(\mathbf{r})/\int p(\mathbf{r})d\mathbf{x},$$

$$\int p(x|y,z)\,dx = 1;$$

$$\left(\frac{\partial f_y}{\partial y}\right) \equiv 2\pi\, p(y|z) = 2\pi[p(y,z)/p(z)] = 2\pi\int p(x,y,z)\,dx / \iint p(x,y,z)\,dx\,dy,$$

$$\int p(y|z)\,dy = 1;$$

$$\left(\frac{\partial f_z}{\partial z}\right) = 2\pi\, p(z) = 2\pi\iint p(x,y,z)\,dx\,dy, \quad \int p(z)\,dz = \int p(\mathbf{r})\,d\mathbf{r} = 1. \quad (103)$$

Indeed, substituting Eq. (103) into Eq. (102) gives

$$\frac{\partial f}{\partial \mathbf{r}} = (2\pi)^3 \left[\frac{p(\mathbf{r})}{p(y,z)}\right]\left[\frac{p(y,z)}{p(z)}\right] p(z)$$

$$= (2\pi)^3 p(x|y,z)\, p(y|z)\, p(z) = (2\pi)^3 p(\mathbf{r}). \quad (104)$$

The HZM determinant of Eq. (82) thus describes N indistinguishable electrons generating the prescribed overall distribution of electrons in the specified molecular system M(N),

$$\rho_M(\mathbf{r}) = Np(\mathbf{r}) = \sum_l |\psi_l(\mathbf{r})|^2 \equiv \sum_l \rho_l(\mathbf{r}), \quad (105)$$

with each of EO densities $\{\rho_l(\mathbf{r})\}$ determining a separate (mutually-closed) subsystem [9, 13] in the combined (constrained) orbital system

$$M_c(N) = (\psi_1 | \psi_2 | \ldots | \psi_N). \quad (106)$$

These "nonbonded" EO subsystems, each comprising a single electron, $\{n_l^+ = 1\}$, thus exhibit separate orbital phases, functionals of corresponding orbital densities (probability distributions): $\{\Phi_l(r) = \Phi[\rho_l; r]\}$ [91, 92].

As argued in Sect. 2 [see Eq. (18)], the externally-open (unconstrained) analog of $M_c(N)$,

$$M_o(N) = (\psi_1 \vdots \psi_2 \vdots \ldots \vdots \psi_N), \tag{107}$$

results from coupling the mutually-closed EO subsystems of $M_c(N)$ to a single "molecular" electron reservoir $\mathcal{R}(\mu)$ in the (macroscopic) composite system $\mathcal{M}_M^* = [\mathcal{R}(\mu) \vdots M_o(N)]$ composed of (mutually-open) EO components in $M_o(N)$:

$$\mathcal{M}_M^* = [\mathcal{R}(\mu) \vdots \psi_1(\mu) \vdots \psi_2(\mu) \vdots \ldots \vdots \psi_N(\mu)] = [\mathcal{R}(\mu) \vdots M_o(N)]. \tag{108}$$

This mutual- and external-opening of the EO subsystems thus implies their effectively "bonded" character, their common phase descriptor of the whole system, $\Phi_M(r) = \Phi[\rho_M; r]$, and fractional orbital populations $\{0 < n_l^* < 1\}$, with finite occupations of the initially empty (virtual) EO states in $M_c(N)$.

Indeed, the externally-open orbital systems must be described by the statistical mixture of EO states $\{|\psi_l\rangle\}$, defined by the associated density operator

$$\sum_l |\psi_l\rangle P_l \langle \psi_l|, \quad \sum_l P_l = 1, \tag{109}$$

with the (external) EO probabilities $\{P_l\}$ determined by the imposed thermodynamic conditions. As representing the mutually-open (effectively bonded, entangled) orbital components this mixed state corresponds to an equalized (molecular) phase intensity and a common level of the chemical potential of the electron reservoir: $\mu = \mu_M$. In the given absolute temperature T of the heat bath $\mathcal{B}$, $T = T_\mathcal{B}$, the equilibrium probability of $\psi_l(\mu)$ subsystem in the EO grand-ensemble is determined by thermodynamic parameters as well as the orbital occupations $\{n_l^*\}$ and energies $\{\varepsilon_l = \langle \psi_l | \hat{H} | \psi_l \rangle\}$,

$$P_l(\mu, T; v) = \Xi^{\mathrm{EO}}(\mu, T; v)^{-1} \exp[\beta(\mu n_l^* - \varepsilon_l)]. \tag{110}$$

Here, $\Xi^{\mathrm{EO}}(\mu, T; v) = \sum_l \exp[\beta(\mu n_l^* - \varepsilon_l)]$ denotes the EO grand-partition function, $\beta = (k_B T)^{-1}$, and k_B is the Boltzmann constant.

These equilibrium probabilities ultimately determine the relevant ensemble averages in this equilibrium "thermodynamic" state, e.g., of the physical observable $\hat{A}$

$$A(\mu, T; v) = \sum_l P_l(\mu, T; v) \langle \psi_l | \hat{A} | \psi_l \rangle = \sum_l P_l(\mu, T; v) A_{l,l} \equiv \langle A \rangle_{ens.}, \tag{111}$$

or the resultant information quantity corresponding to operator $\hat{I}$:

$$I(\mu, T; v) = \sum_l P_l(\mu, T; v) \langle \psi_l | \hat{I} | \psi_l \rangle = \sum_l P_l(\mu, T; v) I_{l,l} \equiv \langle I \rangle_{ens.}. \tag{112}$$

These supplementary physical and IT descriptors of the open EO systems also reflect the external "entangling" ("bonding") of these orbital components.

Each EO component $\psi_l[\rho; r]$ of the orbital system $M_c(N)$ exhibits the phase intensity $F_l(r)$ reflecting the fragment orbital density $\rho_l(r)$, while the equalized phase descriptor is determined by the molecular density $\rho_M(r)$. The effects of a "thermodynamic" opening of these orbital components can be thus unraveled only using the complex orbitals and resultant measures of the "order" (determinicity) or "disorder" (uncertainty) descriptors. Indeed, disentangling (closing) the orbitals in $M_c(N)$ changes the kinetic energy of electrons, by the phase-dependent contribution of Eqs. (36) and (37), thus also affecting the state overall gradient-information content. In other words, the density-alone description is not sufficient to distinguish between such two orbital systems.

As a final comment let us again return to the phase-evolution relation [Eqs. (31) and (38)]. Consider a general partitioning of a molecule M into fragments $\{M_\alpha\}$, mutually closed in $M_c = = (\alpha_1^+ | \alpha_2^+ | \ldots)$ and open in $M_o = (\alpha_1^* | \alpha_2^* | \ldots)$, which yield the same molecular electron density $\rho_M(r)$ in both these composite systems: $\rho_M(r) = \sum_\alpha \rho_\alpha^+(r) = \sum_\alpha \rho_\alpha^*(r)$. Such hypothetical states of subsystems differ only in their phase descriptors, which generate different overall contents of the electronic state entropy or information. One observes that a mutual opening of molecular fragments in $\mathbf{M}_c$, which yields the equilibrium molecular state in $\mathbf{M}_o$, does not affect both the modulus ($\sigma_\phi[R]$) and the external potential ($\sigma_\phi[v]$) contributions to the local phase-production of Eq. (39). Only the remaining, phase-related contribution $\sigma_\phi[\phi]$ to phase-source derivative determines its entanglement displacement in $\mathbf{M}_o$ compared to $\mathbf{M}_c$,

$$\Delta\sigma_\phi = \sigma_\phi[\mathbf{M}_o] - \sigma_\phi[\mathbf{M}_c] = \sigma_\phi[\phi, \mathbf{M}_o] - \sigma_\phi[\phi, \mathbf{M}_c]$$
$$= \sigma_\phi[\phi, \mathbf{M}_o] - \sum_\alpha \sigma_\phi[\phi_\alpha, \mathbf{M}_\alpha] \equiv \sigma_\phi^{nadd.}[\phi]. \qquad (113)$$

It is seen to measure the nonadditive component $\sigma_\phi^{nadd.}[\phi]$ of this nonclassical phase-production equation.

In the Appendix we have additionally examined the one-dimensional particle-in-the-box problem involving the opaque division wall modelling a gradual opening of the partition "left" and "right" subsystems. Attributing the equilibrium phases to EO reconstructing these probability pieces provides the wavefunction description of the mutually-closed (nonbonded, disentangled) subsystems, while the equilibrium phase of the partially open box as a whole corresponds to the gradually-open (bonded, entangled) case.

9. CHEMICAL POTENTIAL AND PHASE-DEPENDENCE OF ELECTRONIC ENERGY

The (local) phase component of molecular wavefunctions for the specified electron density influences the state average kinetic energy contribution to the system overall electronic energy, which also reflects its resultant information content [see Eqs. (35) and (36)]. The intriguing question then arises: how does this presence of the phase-dependent term affect the electronic chemical potential of Eq. (5), measuring the populational (N) derivative of the ensemble-average energy for the fixed external potential v due to the system "frozen" nuclei?

It follows from the Hohenberg-Kohn theorems [40] of modern DFT [40-45] that the ground-state energy $E_0 \equiv E[N, v]$ represents a functional of the ground-state density $\rho[N, v] \equiv N p[N, v] \equiv \rho$ [see also Eq. (2)]:

$$E_v[\rho] = E_v[\rho[N, v]] = E[N[\rho], v[\rho]] \equiv E[\rho].$$

Indeed, this electron distribution identifies uniquely the parameters of the system BO Hamiltonian $\hat{H}(N, v)$, overall number of electrons $N = \int \rho \, d\mathbf{r} \equiv N[\rho]$ and (shape) of the external potential $v = v[\rho]$. It thus also identifies its lowest (nondegenerate) eigensolution

$$\Psi[N, v] = \Psi[N[\rho], v[\rho]] \equiv \Psi[\rho].$$

As argued by Ayers [93] the $\rho \rightarrow v$ mapping is in fact effected by the equilibrium probability distribution $p = \rho/N$ alone, thus implying the (reversible) mappings $v = v[p]$ and $p = p[v]$. Therefore, the constraint of the fixed v can be alternatively effected by fixing the state density-per-particle:

$$\mu[N, v] = (\partial E[N, v]/\partial N)_v = (\partial E[N, v[p]]/\partial N)_p = \mu[\rho] \equiv \mu[Np]. \qquad (114)$$

Consider now the independent modulus (M) and phase (Φ) components of the system complex wavefunction $\Psi[M, \Phi] \equiv M \exp(i\Phi)$. The modulus part of this quantum state ultimately determines the electron density and hence also its normalization to the overall number of electrons: $M \rightarrow N$ or $N = N[M]$. The phase part similarly identifies the system external potential (geometry) [88], $\Phi \rightarrow v$ or $v = v[\Phi]$, and SE implies the reverse mapping $v \rightarrow \Phi$ or $\Phi = \Phi[v]$. Therefore, fixing the external potential in the partial differentiation of the preceding equation is synonymous with fixing the phase part of the electronic wavefunction. This chemical potential (negative electronegativity) derivative thus excludes any nonclassical contribution to this DFT reactivity descriptor, despite a presence of the phase-dependent contribution in the average kinetic energy of electrons, which also reflects the state resultant gradient information.

Consider the *grand* ensemble representing the *externally*-open molecule $M^*(\mu_R; v)$ (see Figure 1c) at the zero temperature limit, $T \to 0$, coupled to electron reservoir $\mathcal{R}(\mu_R)$ in the combined (macroscopic) system $\mathcal{M}(\mu_R) = [\mathcal{R}(\mu_R) \mid M^*(\mu_R)]$ (Figure 2b). This mixture of molecular *ground*-states $\{\psi_0^i = \psi_0[N_i \equiv i, v]\}$, defined for the integer number of electrons $\{N_i\}$ and corresponding to energies $\{E_0^i = E_0[i, v]\}$, which appear in the ensemble with equilibrium thermodynamic probabilities $\{P_0^i(\mu, T \to 0; v)\}$, represents the lowest equilibrium state of an externally open molecule $M^*(\mu, T \to 0; v)$ in such (imposed externally) thermodynamic conditions. In this temperature limit only two ground ($j = 0$) states, ψ_0^i and ψ_0^{i+1}, corresponding to the neighboring integers "bracketing" the given (fractional) average number of electrons $\mathcal{N}$,

$$N_M^* = i P_i(T \to 0) + (i+1)[1 - P_i(T \to 0)] = \mathcal{N}, \quad i \leq \mathcal{N} \leq i+1, \quad (115)$$

appear in the lowest equilibrium *mixed* state [79, 80]. Their ensemble probabilities for the specified average number of electrons $\mathcal{N}$ then read:

$$P_i(T \to 0) = 1 + i - \mathcal{N} \equiv 1 - \omega \quad \text{and} \quad P_{i+1}(T \to 0) = \mathcal{N} - i \equiv \omega. \quad (116)$$

The continuous energy function $E(\mathcal{N}, v)$ then consists of the *straight-line* segments between the neighboring integer values of $\mathcal{N}$. This implies constant values of the chemical potential in all such admissible (partial) ranges of the system average number of electrons and μ-discontinuity at the integer values of the average electron-number, e.g., for $\mathcal{N} = i$, when $\omega = 0$, i.e., $P_i(T \to 0) = 1$, and $P_{i+1}(T \to 0) = 0$ [80]. This pure electronic state, for which the ensemble entropy of von Neumann [94] identically vanishes, indeed corresponds to the equilibrium (ground) state of the neutral molecule as a whole: $M^*(\mu_R; v) = M(\mu_R; v)$.

Conclusion

The classical *modulus*-variable of electronic wavefunction determines the particle probability distribution in the specified molecular state - the electronic structure of "being", while the gradient of the nonclassical *phase*-variable generates the state current which reflects the electronic structure of "becoming". They are both related by the quantum continuity relations implied by molecular SE. The optimum classical degree-of-freedom of molecular states, their density or probability distributions, is determined by the variational principle for the system electronic energy, while the equilibrium shapes of nonclassical state parameters, of the state phase or its current, result from subsidiary extremum rules for the resultant entropy/information measures. The proportionality relation between the state resultant gradient-information and its average kinetic energy of electrons then allows one to interpret the thermodynamic principles as equivalent information criteria, and to use molecular virial theorem in general reactivity considerations.

The overall measures of the information/entropy content in quantum electronic states of molecules and their constituent fragments also open new interpretative possibilities in the chemical reactivity theory and thermodynamic-like description of general molecular systems. In such phenomenological approaches it is customary to distinguish the mutually bonded (open, entangled) and nonbonded (closed, disentangled) status of the given set of electronic distributions in subsystems. It has been argued that the bonded or nonbonded/frozen status of molecular fragments at these hypothetical reaction stages determines specific categories of electronic communications between them. Thus, the promolecular reference state of R^0 does not allow any communications between AIM or their basis functions, the polarized reactive system R^+ opens internal communications within each reactant, while the fully "relaxed" molecular system $R = R^*$ involves all *intra-* and *inter-*substrate propagations.

We have demonstrated in this analysis that the phase components of electronic wavefunctions play a crucial role in a coherent IT treatment of the donor and acceptor reactants. It has been explicitly demonstrated using the additive resultant measures of the entropy/information content in quantum electronic states that the *bonded* status of molecular fragments implies the *phase-equalization* in reactants, at the molecular-phase distribution. Therefore, the bonding (entangling) of molecular fragments represents the *phase*-phenomenon of their common electronic state. The phase-equalization is independent of the actual distance between subsystems, thus representing a long-range correlation effect. This IT-based *phase*-criterion of electronic equilibria in molecular systems complements the familiar energetic principle of the chemical-potential (electronegativity) equalization. The complete set of requirements for the quantum equilibrium in bonded reactive system thus calls for equalizations of both the modulus (probability) and phase (current) related (local) intensities, the chemical potentials and phases of molecular substrates, at the corresponding global distributions characterizing the reactive system as a whole.

As an illustration we have also examined the HZM construction of molecular wavefunctions in molecules, which yields the prescribed electron distribution. The EO states determining the system Slater determinant define the constrained multicomponent system, composed of the mutually-closed orbital units, with each subsystem being characterized by its own phase and chemical-potential descriptors. Their simultaneous opening onto a common electron reservoir, and hence also onto themselves, generates the externally- and mutually-open orbital system in which these EO subsystems are effectively "bonded" (entangled). They then exhibit a common phase descriptor and equalize their chemical potentials at the reservoir level. This mixed equilibrium state is determined by the density operator corresponding to thermodynamic (grand-ensemble) probabilities of EO, related to orbital energies and their average occupations.

The overall (additive) descriptor of the gradient-information content in the specified information-equilibrium electron configuration, the Slater determinant constructed from the occupied EO of Eq. (96), is determined by the sum $I[p, \psi_k^{eq.}[p]] = \sum_l I[p, \psi_l^{eq.}[p]]$ of the occupied EO contributions

$$I[p, \psi_l^{eq.}[p]] = I[p] + 4(\delta k_l[p])^2 \int p(r)[\nabla \cdot f[p; r]]^2 dr. \qquad (117)$$

The minimum value of such resultant gradient information for the prescribed probability distribution $p(r)$ is thus reached for $\{\delta k_l^{eq.}[p] = \mathbf{0}\}$, or $\{k_l^{eq.}[p] = \langle k[p]\rangle\}$, i.e., for the highest EO uncertainty (entropy), i.e., the lowest orbital determinicity (information).

Let us finally comment on the EO phase-consistency relation of Eq. (27). For a single electron the orthogonality constraints do not intervene, $F_1(r) = const.$, so that $\nabla F_1(r) = \mathbf{0}$ and $\Delta F_1(r) = 0$. In *many*-electron configurations a single EO $\psi_l[p]$ or its equilibrium analog $\psi_l^{eq.}[p]$ do not satisfy this phase requirement, since the local gradients,

$$\nabla F_l(r) = k_l \nabla \cdot f(r) = 2\pi k_l P(U) \neq 0 \quad \text{or}$$

$$\nabla \Phi_l(r) = k_l \nabla \cdot f(r) + \nabla \phi(r) \neq 0, \tag{118}$$

give rise to a nonvanishing phase-Laplacian of each occupied EO.

Consider now the electron configurations of Eq. (82) or its equilibrium analog

$$\Psi_{\mathbf{k}}^{eq.}[p] = |\psi_1^{eq.}[p]\,\psi_2^{eq.}[p]\ldots\psi_N^{eq.}[p]| \equiv \det(\psi_{\mathbf{k}}^{eq.}[p]). \tag{119}$$

For the *N*-particle "coalescence" in r,

$$r_1 = r_2 = \ldots = r_N = r, \tag{120}$$

these *N*-electron wavefunctions exhibit the resultant phases expressed by sums over phases of the occupied EO:

$$\Phi_{\mathbf{k}}(N; r) = \sum_l F_l(r) = \sum_l k_l \cdot f(r) \quad \text{or}$$

$$\Phi_{\mathbf{k}}^{eq.}(N; r) = \Phi_{\mathbf{k}}(N; r) + N\phi(r) = \sum_l \delta k_l \cdot f(r). \tag{121}$$

The phase requirement in $\Psi_{\mathbf{k}}^{eq.}[p]$, $\Delta\Phi_{\mathbf{k}}(N; r) = 0$, imposes the constraint $K(\mathbf{k}) \equiv \sum_l k_l = \mathbf{0}$, on the EO wave-vectors $\{k_l\}$, which then implies the vanishing values of the phase gradient, $\nabla\Phi_{\mathbf{k}}(N; r) = 2\pi K(\mathbf{k}) P(U) = \mathbf{0}$, and phase Laplacian, $\nabla^2 \Phi_k(N; r) = 0$. In the equilibrium configuration δk_l replaces k_l, so that the phase requirement is automatically satisfied,

$$\nabla \Phi_k^{eq.}(N; r) = 2\pi \, \delta K(\mathbf{k}) \, P(U) = 0, \tag{122}$$

since $\delta K(\mathbf{k}) \equiv \sum_l \delta k_l = 0$.

APPENDIX: MODELLING SUBSYSTEM OPENING WITH OPAQUE DIVISION WALL

Consider the one-dimensional (x) wave equation for a single particle enclosed in a box of impenetrable potential walls at $x = -a$ and $x = +a$, divided by an infinitely high opaque wall at $x = 0$, of infinitely small width. It can serve as the solvable problem modelling the mutual opening of two "left" and "right" subsystems. The dividing potential wall can be then analytically described by the Dirac δ–function,

$$v(x) = (\hbar^2/m) \, \Omega \, \delta(x), \tag{A1}$$

controlled by the (positive) *opacity* parameter Ω. The larger Ω, the more opaque the wall becomes: $\Omega = 0$ represents a fully transparent (permeable) wall " ¦ " in the mutually-*open* subsystems, the limit $\Omega \to \infty$ corresponds to the opaque (impermeable) wall " | " separating mutually-*closed* subsystems, while a finite opacity implies a controlled partial penetration of the wall. In other words, the impermeable wall separates two potential subboxes of width a (separate "subsystems"), called a-boxes, while the transparent case gives rise to a single potential box of a double width, called $2a$-box. A detailed discussion of this stationary SE problem,

$$d^2u(x)/dx(x)^2 + [k^2 - 2\Omega\delta(x)]u(x) = 0, \quad k^2 = 2mE/\hbar^2, \tag{A2}$$

for determining the system wavefunctions (Hamiltonian eigenstates)

$\{\psi_n(x,t) = u_n(x)\exp[-i(E_n/\hbar)t]\}$

and the associated energy levels (Hamiltonian eigenvalues)

$\{E_n = (k_n\hbar)^2/(2m)\}, \quad n = 1, 2, 3, \ldots,$

can be found in ref. [95].

Since this differential equation and boundary conditions $u(\pm a) = 0$ are invariant with respect to parity transformation $x \to -x$, its eigenfunctions can be classified as being either of even $[u_n^+(x)]$ or odd $[u_n^-(x)]$ parity:

$u_n^+(x) = u_n^+(-x) \quad$ and $\quad u_n^-(x) = -u_n^-(-x).$

Examining the neighbourhood of the potential wall and integrating SE across the wall, u being continuous and prime denoting the spatial differentiation, then implies the derivative discontinuity

$$u'(+0) - u'(-0) = 2\Omega u(0), \tag{A3}$$

which identically vanishes for the odd eigenstates, for which $u_n^-(0) = 0$. Such states thus exhibit a continuous derivative without any jump at the wall. Therefore, the odd solutions are unaffected by the wall presence, however opaque.

For the even eigenstates, which exhibit a finite amplitude at $x = 0$, one observes at the wall a break in the state curve and a jump of 2Ω in its logarithmic derivative $L[u] = [\ln u]' = u'/u$:

$$L[u_n^+(+0)] - L[u_n^+(-0)] = 2\Omega. \tag{A4}$$

Indeed, the eigenfunctions of even parity, of a general form

$$u_n^+(x) = \{-A\sin[k_n^+(x+a)] \quad \text{for} \quad -a \le x < 0,$$
$$\phantom{u_n^+(x) = \{}A\sin[k_n^+(x-a)] \quad \text{for} \quad 0 < x \le a\}, \tag{A5}$$

predict

$$L[u_n^+(+0)] = -k_n^+\cot(k_n^+ a), \quad L[u_n^+(-0)] = +k_n^+\cot(k_n^+ a)\},$$

and hence, from Eq. (A4),

$$k_n^+\cot(k_n^+ a) = -\Omega.$$

The lowest symmetric state in the fully transparent model ($\Omega = 0$) thus marks the lowest (symmetric) probability distribution of the double box: $p(x; \Omega = 0) = p_0[2a; x]$, while the corresponding density in the completely opaque case ($\Omega \to \infty$) represents a collection of the two ground-state probability densities for the separate subboxes,

$$\{p_\alpha(x; \Omega \to \infty) = p_0[a; x], \quad \alpha = \text{right, left}\}.$$

Only these extreme stationary distributions correspond to the pure quantum states exhibiting the vanishing local phase.

The subsystem pieces of the equilibrium ("molecular") probability density in the partially opaque systems (for a finite $0 \leq \Omega$), obtained by the spatial division at the wall, are not solutions of SE, since they reflect only a part of the overall influence of the dividing wall in the system as a whole. In particular, for $\Omega = 0$, the left and right halves of $p_0[2a; x]$ should be used as subsystem distributions in the closed (c) system

$$M_c = [\text{left}(\Omega = 0) | \text{right}(\Omega = 0)],$$

to determine phase changes accompanying their mutual opening in the open (o) system

$$M_o = [\text{left}(\Omega = 0) ¦ \text{right}(\Omega = 0)].$$

As we have argued in Section 8, the density pieces $\{p_\alpha(x; \Omega)\}$ can be still associated with the (local) equilibrium phases $\phi[p_\alpha; r]$ [Eq. (97)] attributed to EO reconstructing these known subsystem distributions. A similar procedure can be adopted for any finite value of the opacity parameter Ω, reflecting a gradually modified transparency of the dividing wall.

REFERENCES

[1] Fisher RA (1925) Theory of statistical estimation. *Proc Cambridge Phil Soc* 22:700-725.

[2] Frieden BR (2004) *Physics from the Fisher information – a unification*. Cambridge University Press, Cambridge.

[3] Shannon CE (1948) The mathematical theory of communication. *Bell System Tech J* 7:379-493, 623-656.

[4] Shannon CE, Weaver W (1949) *The mathematical theory of communication*. University of Illinois, Urbana.

[5] Kullback S, Leibler RA (1951) On information and sufficiency. *Ann Math Stat* 22:79-86.

[6] Kullback S (1959) *Information theory and statistics*. Wiley, New York.

[7] Abramson N (1963) *Information theory and coding*. McGraw-Hill, New York.

[8] Pfeifer PE (1978) *Concepts of probability theory*. Dover, New York.

[9] Nalewajski RF (2006) *Information theory of molecular systems*. Elsevier, Amsterdam.

[10] Nalewajski RF (2010) *Information origins of the chemical bond*. Nova Science Publishers, New York.

[11] Nalewajski RF (2012) *Perspectives in electronic structure theory*. Springer, Heidelberg.

[12] Nalewajski RF, Parr RG (2000) Information theory, atoms-in-molecules and molecular similarity. *Proc Natl Acad Sci USA* 97:8879-8882.

[13] Nalewajski RF (2003) Information principles in the theory of electronic structure. *Chem Phys Lett* 272:28-34.
[14] Nalewajski RF (2003) Information principles in the Loge Theory. *Chem Phys Lett* 375:196-203.
[15] Nalewajski RF, Broniatowska E (2003) Information distance approach to Hammond Postulate. *Chem Phys Lett* 376:33-39.
[16] Nalewajski RF, Parr RG (2001) Information-theoretic thermodynamics of molecules and their Hirshfeld fragments. *J Phys Chem A* 105:7391-7400.
[17] Nalewajski RF (2002) Hirschfeld analysis of molecular densities: subsystem probabilities and charge sensitivities. *Phys Chem Chem Phys* 4:1710-1721.
[18] Parr RG, Ayers PW, Nalewajski RF (2005) What is an atom in a molecule? *J Phys Chem A* 109:3957-3959.
[19] Nalewajski RF, Broniatowska E (2007) Atoms-in-Molecules from the stockholder partition of molecular two-electron distribution. *Theoret Chem Acc* 117:7-27.
[20] Heidar-Zadeh F, Ayers PW, Verstraelen T, Vinogradov I, Vöhringer-Martinez E, Bultinck P (2018) Information-theoretic approaches to Atoms-in-Molecules: Hirshfeld family of partitioning schemes. *J Phys Chem A* 122:4219-4245.
[21] Hirshfeld FL (1977) Bonded-atom fragments for describing molecular charge densities. *Theoret Chim Acta* (Berl) 44:129-138.
[22] Nalewajski RF (2000) Entropic measures of bond multiplicity from the information theory. *J Phys Chem A* 104:11940-11951.
[23] Nalewajski RF (2004) Entropy descriptors of the chemical bond in Information Theory: I. Basic concepts and relations. *Mol Phys* 102:531-546; II. Application to simple orbital models. *Mol Phys* 102:547-566.
[24] Nalewajski RF (2004) Entropic and difference bond multiplicities from the two-electron probabilities in orbital resolution. *Chem Phys Lett* 386:265-271.

[25] Nalewajski RF (2005) Reduced communication channels of molecular fragments and their entropy/information bond indices. *Theoret Chem Acc* 114:4-18.

[26] Nalewajski RF (2005) Partial communication channels of molecular fragments and their entropy/information indices. *Mol Phys* 103:451-470.

[27] Nalewajski RF (2011) Entropy/information descriptors of the chemical bond revisited. *J Math Chem* 49:2308-2329.

[28] Nalewajski RF (2014) Quantum information descriptors and communications in molecules. *J Math Chem* 52:1292-1323.

[29] Nalewajski RF (2009) Multiple, localized and delocalized/conjugated bonds in the orbital-communication theory of molecular systems. *Adv Quant Chem* 56:217-250.

[30] Nalewajski RF, Szczepanik D, Mrozek J (2011) Bond differentiation and orbital decoupling in the orbital communication theory of the chemical bond. *Adv Quant Chem* 61:1-48.

[31] Nalewajski RF, Szczepanik D, Mrozek J (2012) Basis set dependence of molecular information channels and their entropic bond descriptors. *J Math Chem* 50:1437-1457.

[32] Nalewajski RF (2017) Electron communications and chemical bonds. In *Frontiers of quantum chemistry*, Wójcik M, Nakatsuji H, Kirtman B, Ozaki Y (Eds). Springer, Singapore, pp 315-351.

[33] Nalewajski RF, Świtka E, Michalak A (2002) Information distance analysis of molecular electron densities. *Int J Quantum Chem* 87:198-213.

[34] Nalewajski RF, Broniatowska E (2003) Entropy displacement analysis of electron distributions in molecules and their Hirshfeld atoms. *J Phys Chem A* 107:6270-6280.

[35] Nalewajski RF (2008) Use of Fisher information in quantum chemistry. *Int J Quantum Chem* (Jankowski K issue) 108:2230-2252.

[36] Nalewajski RF, Köster AM, Escalante S (2005) Electron localization function as information measure. *J Phys Chem A* 109:10038-10043.

[37] Becke AD, Edgecombe KE (1990) A simple measure of electron localization in atomic and molecular systems. *J Chem Phys* 92:5397-5403.
[38] Silvi B, Savin A (1994) Classification of chemical bonds based on topological analysis of electron localization functions. *Nature* 371:683-686.
[39] Savin A, Nesper R, Wengert S, Fässler TF (1997) ELF: the electron localization function. *Angew Chem Int Ed Engl* 36:1808-1832.
[40] Hohenberg P, Kohn W ((1964) Inhomogeneous electron gas. *Phys Rev* 136B:864-971.
[41] Kohn W, Sham LJ (1965) Self-consistent equations including exchange and correlation effects. *Phys Rev* 140A:133-1138.
[42] Levy M (1979) Universal variational functionals of electron densities, first-order density matrices, and natural spin-orbitals and solution of the v-representability problem. *Proc Natl Acad Sci USA* 76:6062-6065.
[43] Parr RG, Yang W (1989) *Density-functional theory of atoms and molecules*. Oxford University Press, New York.
[44] Dreizler RM, Gross EKU (1990) *Density functional theory: an approach to the quantum many-body problem*. Springer, Berlin.
[45] Nalewajski, RF (Ed) (1996) *Density functional theory I-IV, Topics in Current Chemistry* vols 180-183.
[46] Nalewajski RF, de Silva P, Mrozek J (2010) Use of nonadditive Fisher information in probing the chemical bonds. *J Mol Struct: THEOCHEM* 954:57-74.
[47] Nalewajski RF (2011) Through-space and through-bridge components of chemical bonds. *J Math Chem* 49:371-392.
[48] Nalewajski RF (2011) Chemical bonds from through-bridge orbital communications in prototype molecular systems. *J Math Chem* 49:546-561.
[49] Nalewajski RF (2011) On interference of orbital communications in molecular systems. *J Math Chem* 49:806-815.
[50] Nalewajski RF, Gurdek P (2011) On the implicit bond-dependency origins of bridge interactions. *J Math Chem* 49:1226-1237.

[51] Nalewajski RF (2012) Direct (through-space) and indirect (through-bridge) components of molecular bond multiplicities. *Int J Quantum Chem* 112:2355-2370.

[52] Nalewajski RF, Gurdek P (2012) Bond-order and entropic probes of the chemical bonds. *Struct Chem* 23:1383-1398.

[53] Nalewajski RF (2013) Exploring molecular equilibria using quantum information measures. *Ann Phys* (Leipzig) 525:256-268.

[54] Nalewajski RF (2014) On phase equilibria in molecules. *J Math Chem* 52:588-612.

[55] Nalewajski RF (2014) Quantum information approach to electronic equilibria: molecular fragments and elements of non-equilibrium thermodynamic description. *J Math Chem* 52:1921-1948.

[56] Nalewajski RF (2015) Phase/current information descriptors and equilibrium states in molecules. *Int J Quantum Chem* 115:1274-1288.

[57] Nalewajski RF (2015) Quantum information measures and molecular phase equilibria. In *Advances in mathematics research* vol 19, Baswell AR (Ed). Nova Science Publishers, New York, pp. 53-86.

[58] Nalewajski RF (2014) On phase/current components of entropy/information descriptors of molecular states. *Mol Phys* 112:2587-2601.

[59] Nalewajski RF (2016) Complex entropy and resultant information measures. *J Math Chem* 54:1777-1782.

[60] Nalewajski RF (2018) Phase description of reactive systems. In *Conceptual density functional theory*, Islam N, Kaya S (Eds). Apple Academic Press, Waretown, pp. 217-249.

[61] Nalewajski RF (2017) Quantum information measures and their use in chemistry. *Current Phys Chem* 7:94-117.

[62] Nalewajski RF (2016) *Quantum information theory of molecular states*. Nova Science Publishers, New York.

[63] Nalewajski RF (1980) Virial theorem implications for the minimum energy reaction paths. *Chem Phys* 50:127-136.

[64] Nalewajski RF (2019) Understanding electronic structure and chemical reactivity: quantum-information perspective. In *The*

application of quantum mechanics to the reactivity of molecules, Sousa S (Ed), *Appl Sci* 9:1262-1292.

[65] Nalewajski RF (2019) On entropy/information description of reactivity phenomena. In *Advances in mathematics research* vol 26, Baswell AR (Ed). Nova Science Publishers, New York, pp 97-157.

[66] Nalewajski RF (2019) Role of electronic kinetic energy (resultant gradient information) in chemical reactivity. *J Mol Model* (Latajka Z issue), Berski S, Sokalski WA (Eds) 25:259-278.

[67] Nalewajski RF (2019) On classical and quantum entropy/information descriptors of molecular electronic states. In *Research methodologies and practical applications of chemistry*, Pogliani L, Haghi AK, Islam N (Eds). Apple Academic Press, Waretown, in press.

[68] Nalewajski RF (2018) Information equilibria, subsystem entanglement and dynamics of overall entropic descriptors of molecular electronic structure. *J Mol Model* (Chattaraj PK issue), Sarkar U (Ed) 24:212-227.

[69] Nalewajski RF (2016) On entangled states of molecular fragments. *Trends in Physical Chemistry* 16:71-85.

[70] Callen HB (1962) *Thermodynamics: an introduction to the physical theories of equilibrium thermostatics and irreversible thermodynamics*. Wiley, New York.

[71] Nalewajski RF, Korchowiec J, Michalak A (1996) Reactivity criteria in charge sensitivity analysis. In *Density functional theory IV*, Nalewajski RF (Ed). *Topics in Current Chemistry* 183:25-141.

[72] Nalewajski RF, Korchowiec J (1997) *Charge sensitivity approach to electronic structure and chemical reactivity*. World Scientific, Singapore.

[73] Geerlings P, De Proft F, Langenaeker W (2003) Conceptual density functional theory. *Chem Rev* 103:1793-1873.

[74] Nalewajski RF (1994) Sensitivity analysis of charge transfer systems: *in situ* quantities, intersecting state model and its implications. *Int J Quantum Chem* 49:675-703.

[75] Nalewajski RF (1995) Charge sensitivity analysis as diagnostic tool for predicting trends in chemical reactivity. In *Proceedings of the*

NATO ASI on Density Functional Theory (Il Ciocco, 1993), Dreizler RM, Gross EKU (Eds). Plenum, New York, pp 339-389.

[76] Chattaraj PK (Ed) (2009) *Chemical reactivity theory: a density functional view*. CRC Press, Boca Raton.

[77] Gatti C, Macchi P (Eds) (2012) *Modern charge-density analysis*. Springer, Berlin.

[78] Nalewajski RF (2017) Chemical reactivity description in density-functional and information theories. In *Chemical concepts from density functional theory*, Liu S (Ed). *Acta Phys-Chim Sin* 33:2491-2509.

[79] Gyftopoulos EP, Hatsopoulos GN (1965) Quantum-thermodynamic definition of electronegativity. *Proc Natl Acad Sci USA* 60:786-793.

[80] Perdew JP, Parr RG, Levy M, Balduz JL (1982) Density functional theory for fractional particle number: derivative discontinuities of the energy. *Phys Rev Lett* 49:1691-1694.

[81] Mulliken RS (1934) A new electronegativity scale: together with data on valence states and on ionization potentials and electron affinities. *J Chem Phys* 2:782-793.

[82] Iczkowski RP, Margrave JL (1961) Electronegativity. *J Am Chem Soc* 83:3547-3551.

[83] Parr RG, Donnelly RA, Levy M, Palke WE (1978) Electronegativity: the density functional viewpoint. *J Chem Phys* 69:4431-4439.

[84] Parr RG, Pearson RG (1983) Absolute hardness: companion parameter to absolute electronegativity. *J Am Chem Soc* 105:7512-7516.

[85] Parr RG, Yang W (1984) Density functional approach to the frontier-electron theory of chemical reactivity. *J Am Chem Soc* 106:4049-4050.

[86] Prigogine I (1980) *From being to becoming: time and complexity in the physical sciences*. Freeman WH & Co, San Francisco.

[87] von Weizsäcker CF (1935) Zur theorie der kernmassen. *Z Phys* 96:431-458.

[88] Nalewajski RF (2020) Overall entropy/information descriptors of electronic states and chemical reactivity. In *Mathematics applied to*

engineering: advanced theories, methods, and models, Islam N, Bir Singh S, Ranjan P, Haghi AK (Eds), in press.

[89] Harriman JE (1980) Orthonormal orbitals for the representation of an arbitrary density. *Phys Rev A* 24:680-682.

[90] Zumbach G, Maschke K (1983) New approach to the calculation of density functionals. *Phys Rev A* 28:544-554; Erratum, *Phys Rev A*29:1585-1587.

[91] Nalewajski RF (2019) Equidensity orbitals in resultant-information description of electronic states. *Theoret Chem Acc* 138:108-123.

[92] Nalewajski RF (2019) Resultant information description of electronic states and chemical processes. *J Phys Chem A* (Geerlings P issue) 123:45-60.

[93] Ayers PW (2000) Density per particle as a descriptor of Coulomb systems. *Proc Natl Acad Sci USA* 97:1959-1964.

[94] von Neumann J (1955) *Mathematical foundations of quantum mechanics*. Princeton University Press, Princeton.

[95] Flügge S (1974) *Practical Quantum Mechanics*. Springer-Verlag, Heidelberg.

In: An Introduction to Electronic Structure … ISBN: 978-1-53618-411-2
Editor: Nadia T. Paulsen © 2020 Nova Science Publishers, Inc.

Chapter 2

ELECTRONIC STRUCTURE OF SOLID OXIDES DOPED WITH TRANSITION ELEMENTS

N. Chezhina[*] *and D. Korolev*
Saint Petersburg State University,
Saint Petersburg, Russia

ABSTRACT

The chapter is devoted to the description of the most efficient method for studying the electronic structure of solids – the magnetic dilution method, e.g., the study of magnetic susceptibility of diluted solid solution of paramagnetic oxides in diamagnetic isomorphous matrices. The method gives the possibility to determine the electronic state of a single paramagnetic atom, the energetics of magnetic exchange between paramagnetic atoms, and the character of paramagnetic atoms distribution in a diamagnetic matrix. The influence of the composition of a diamagnetic matrix can be traced and explained. Complex oxides with perovskite, layered K_2NiF_4 type, and spinel structure are considered as the examples of the magnetic dilution method application.

---

[*] Corresponding Author's Email: chezhina.natalia@gmail.com.

Keywords: magnetic susceptibility, magnetic dilution, perovskite, layered structure, spinels

INTRODUCTION

Complex oxides lay the foundation for a large series of materials for energy saving electron-ionic conductors, solar cells, magnetoresistors, catalysts, and so on. Their wide application is based on thermal stability and rather simple preparation. They are used mostly as solid solutions with transition and rare earth elements, which determine their unique properties. A large series of structures, especially perovskite like, are tolerant to substitution varying both by the nature of the substituent and by their concentration. Taking into account the fact that in the Periodic table of elements the number of possible substituents in the metal sublattice amounts to about a hundred, and sometimes several elements are introduced simultaneously, the more so, the ratio between several substituents may vary in wide regions, the number of possible compositions tends to infinity. But the selection of compositions for such complex oxide systems is usually carried out purely experimentally. This results in unjustified expenditures of time and starting materials. Only a thorough study of the electronic structure of such systems would make possible the optimization of qualitative and quantitative composition for obtaining the designated properties.

Electronic structure includes:

1. Valence and spin states of transition element atoms, and the influence of crystal structure and the nature of diamagnetic atoms on these characteristics.
2. The distribution of paramagnetic atoms over various nonequivalent crystallographic sites of the structure.
3. Interatomic interactions in the structure and the influence of the distortions of the nearest surrounding and of diamagnetic atoms in the solid oxide on them.

In other words with the aim of laying off the theoretical grounds for the pathway to selecting the composition of ceramics with predetermined physical and chemical characteristics we must solve the problems thoroughly connected with each other in various ways:

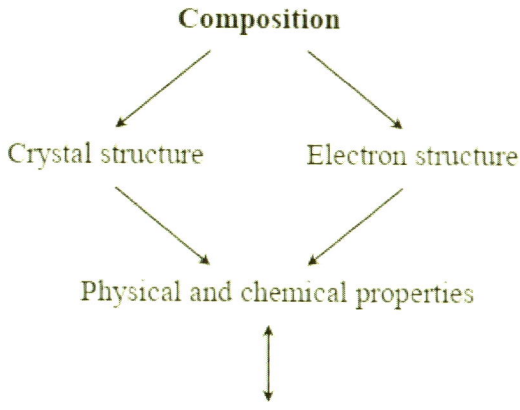

The interrelation between the composition (both qualitative and quantitative) and crystal structure is successfully studied by X-ray methods exquisitely developed nowadays and allowing the structure refinement to be made for powder samples. The problem of electron structure and its interrelation with crystal structure and physicochemical properties is more complicated and can be solved by a number of physical methods sensitive to the state of paramagnetic atoms and interatomic interactions. All of them, such as ESR, NMR, visual spectra, saying nothing about such an informative method as Moessbauer spectroscopy, have special features when applied to solid state and various limits of their application associated with the nature of the elements under study, their valence and spin states. When selecting the compositions of new target materials the scientists must take into account general regularities existing in the Periodic Table of elements – the variations of properties along the periods and groups of elements. This makes it necessary to study a wide number of complex oxide systems containing various elements within the same period or group. Therefore, the search of a universal method sensitive to crystal and electronic structure but not limited by the nature and electron states of elements becomes urgent.

Such a method can be the magnetic susceptibility, which for solids is used in the form of magnetic dilution. The method of static magnetic susceptibility is almost exhaustively developed for complex compounds, and everybody can get acquainted with its fundamentals from our monograph [1] or from the references given herein. But in this chapter we shall discuss the method of magnetic dilution in more details and give appropriate examples of its use for the study of electronic structure of complex oxides containing d-elements.

MAGNETIC DILUTION METHOD

The reason for the development of magnetic dilution method is that in solids containing paramagnetic elements, which occupy all or the most part of available sites in a lattice we always have a superposition of several effects – the state of elements (valence and spin) and interatomic interactions. Those latter may occur in various directions and differ in energy. In the rock salt structure, for example, we can find three possible exchange interactions between atoms in the octahedral surrounding, in the spinel structure there also can be the interaction between atoms occupying oxygen octahedra and tetrahedra. All the attempts to derive exhaustive information about the above mentioned parameters of electron structure from magnetic measurement of one sample [2] give only approximate and averaged values.

Magnetic dilution method allows the problems of electronic structure of solids to be separated into two parts – to determine the states of paramagnetic atoms, the symmetry of their surroundings, spin transitions and to estimate the distribution of paramagnetic atoms in a lattice, the energy of magnetic exchange, and interatomic interactions. For the first problem magnetic characteristics of a series of diluted solid solutions of paramagnetic oxide in an isomorph diamagnetic matrix must be extrapolated to zero concentration. Then we deal with a hypothetic situation when one paramagnetic atom appears in the diamagnetic oxide and all the theory of magnetic susceptibility for complex compounds may be used to describe its electronic structure [3, 4]. The run of the dependences of paramagnetic component of magnetic susceptibility and effective magnetic moment on the concentration of paramagnetic element can be used for solving the problem of interatomic interactions.

Since we deal with diluted solutions (for example $LaM_xAl_{1-x}O_3$, where $0 < x < 0.1$) it seems to be correct to assume that there are single paramagnetic atoms in the lattice and small clusters of two-four atoms, which do not interact between each other. Then the paramagnetic component of the susceptibility may be calculated as the sum of these susceptibilities with respect for their fractions:

$$\chi_M^{para} = \sum_i a_i \chi_i \tag{1}$$

where a_i – the fractions of monomers and various clusters, χ_i – their susceptibilities. Those latter for clusters are the functions of the exchange parameter J. The susceptibility of monomers may be taken from the values extrapolated to zero concentration. When the concentration interval is limited by $0 - 0.05$ only clusters of two paramagnetic atoms can be taken into account since the formation of greater clusters is highly improbable. And for such a case formula (1) becomes

$$\chi_M^{para} = a_1 \chi_1 + a_2 \chi_2, \tag{2}$$

which means that we have two variables for each concentration – a_2 ($a_1 = 1 - a_2$) and χ_2, the susceptibility of dimer clusters being a function of J and must be the same for all the dimers if we have only one type of dimers. For example for perovskite oxides they are M-O-M, whereas the fraction of dimers is increasing as the concentration **increases**. To calculate χ_2 we use Heisenberg-Dirac-van Vleck model, which formulae are given elsewhere [1, 4].

The effective magnetic moment of a binuclear cluster is calculated by summing up the squares of separate moments $\mu_{eff}^2(S')$ over all the spin layers with respect to a corresponding Boltzmann's factor. Then the effective magnetic moment of a binuclear cluster is expressed by equation (3)

$$\mu_{eff}^2 = \frac{g^2 \sum_{S'} S'(S'+1)(2S'+1) \cdot \exp[-E(S')/kT]}{n \sum_{S'} (2S'+1) \cdot \exp[-E(S')/kT]} \qquad (3)$$

where $E(S') = -J[S'(S'+1) - S_1(S_1+1) - S_2(S_2+1)]$,

$S' = (S_1 + S_2), (S_1 + S_2 - 1), ..., |S_1 - S_2|$

the total spin of the cluster, g – Lande's factor, n – number of magnetic centers in cluster.

Since we have 6-7 solid solutions in the most diluted region, their susceptibility being measured for 15-20 fixed temperatures in the range 77-400 K, the calculations give a_2 and J with good certainty. After plotting a_2 vs x we can make the estimations of the energy of interatomic interactions.

When comparing the obtained exchange parameters for various d-elements in the same diamagnetic matrix we can use the model of exchange channels [4], which suggests that the total exchange parameter is a sum of exchange parameters along each possible channel – antiferromagnetic or ferromagnetic. The number and character of these channels depends on the electronic structure of a paramagnetic atom.

After plotting a_2 vs x we can make the estimations of the energy of interatomic interactions. From the point of view of thermodynamics according to the data of [5, 6] solid solutions of oxide systems are regular solutions, i.e., the systems with a small, but as a rule nonzero mixing enthalpy, which is related to the interchange energy

$$W_{12} = H_{12} - \tfrac{1}{2}(H_{11} + H_{22}) \qquad (4)$$

where H_{ij} is the energy of interaction between like and unlike atoms in the solid solution.

This is important in the process of considering magnetic properties, since the distribution of the atoms is strictly statistical only in the ideal solutions. Any deviation from the ideal inevitably results in certain segregation. Here is the reason for inadequacy of using the statistical distribution for calculating the susceptibility of magnetically diluted systems. Then, using the equations of statistical thermodynamics [5] and having the plots of paramagnetic atom distribution (a_2 for the whole concentration range under study), we can estimate the interchange energy (5):

$$\frac{\left(x - \frac{a_2 x}{z}\right)^2}{\frac{a_2 x}{z}\left(1 - 2x + \frac{a_2 x}{z}\right)} = \exp\left(-\frac{2W_{12}}{kT}\right) \quad (5)$$

where x is the concentration, a_2 – the fraction of dimers, z – coordination number of d-element, W_{12} – the interchange energy.

To use the magnetic dilution method rigid requirements must be met. First of all the solid solutions under study must be exceptionally pure, devoid of any admixtures. With this purpose in mind the starting materials are to be at least 99.999%, what is more, all the diamagnetic starting oxides must be checked by magnetic susceptibility method. The matter is that even 0.001% of iron(III) would give the susceptibility χ_g of about $+0.3 \cdot 10^{-6}$ emu/g at room temperature and if iron is in Fe_3O_4, the susceptibility will depend on the field strength. This is significant, since, for example, $\chi_g(TiO_2)$ is about $-0.3 \cdot 10^{-6}$ emu/g. The obtained solid solutions must be thoroughly characterized by their structure (X-ray) and composition. The second requirement for the solid solutions is that they must be close to the equilibrium state, since after the system becomes single phase, the distribution of a paramagnetic element may change upon further sintering. Only an additional sintering, which shows the susceptibility remaining constant and even remaining constant after sintering at an elevated temperature and then again at the temperature of the synthesis may assure the researcher that the distribution of the paramagnetic atoms is equilibrium, and thermodynamic estimations will be valid.

The main parameters derived from magnetic susceptibility measurements are the paramagnetic component of magnetic susceptibility calculated per 1 mole of paramagnetic component (χ_M) and the effective magnetic moment (μ_{eff}) calculated from Curie equation

$$\chi_M^{para} = \frac{N\beta^2 \mu_{eff}^2}{3kT} \tag{6}$$

where N is Avogadro number, β – Bohr magneton, k - Boltzmann constant.

When calculating the paramagnetic component of susceptibility of a solid solution it appears impossible to use the tabulated data of diamagnetic increments for the following reasons. First, the total susceptibility of a diluted solid solution may be small in comparison to the sum of diamagnetic corrections, which will result in a substantial error, and second, since it is impossible from thermodynamic point of view to obtain an absolutely pure substance, even the susceptibility of a diamagnetic matrix inevitably will depend on temperature to some extent. Thus, the susceptibility of the diamagnetic matrix must be measured over the same temperature range as the solid solutions and the paramagnetic component of susceptibility is determined by equation (7)

$$\chi_M^{para} = \frac{\chi_g^{sol} M^{sol} - (1-x)\chi_M^{mat} M^{mat}}{x} - \sum_i \chi_i^{dia}, \tag{7}$$

where χ_g^{sol} and χ_g^{mat} are measured specific susceptibilities of the solid solution and of diamagnetic matrix respectively, M^{sol} and M^{mat} - their molar masses, $\sum_i \chi_i^{dia}$ – the sum of tabulated diamagnetic corrections for the dissolved substance.

As is seen from formulae (7), the accuracy of determining χ_M^{para} depends on x in the solid solution formula. The latter can deviate from the predetermined, since the solid solutions usually are obtained from sintering the starting materials at rather high temperatures. Thus, a good chemical analysis for the content of paramagnetic element in the resulting system is necessary.

Therefore, we can see that the use of magnetic dilution method, though informative from the point of view of electronic structure of oxide systems, requires a great preliminary chemical work to obtain reliable results.

In this work we present the results of our investigations demonstrating the possibilities of the method of magnetic dilution for particular oxide systems.

COMPLEX OXIDES WITH PEROVSKITE AND PEROVSKITE-LIKE LAYERED STRUCTURES

The interests of the researchers to complex oxides with perovskite-like structure developed in a wave-like way over the latest 50 years. First ABO_3 perovskites attracted attention as catalysts, later they gave way to complex oxides with K_2NiF_4 structure and a large series of Aurivillius phases, and by the end of eighties perovskites again occupy the major place in the chemistry of oxide materials. This is associated with the necessity to obtain new materials for energy saving technologies. But these materials usually are associated with heterovalent doping of perovskites (magnetoresistors [1, 7], electrodes for solid oxide fuel cells [1, 8]) and layered complex oxides [9, 10]. The problem of describing electronic structure of such systems is even more complicated, but urgent, since the principles of selecting the compositions are still obscure. Therefore, to address the researchers to the problem of heterovalent doping of complex oxides, which we successfully solved for the last decades [11, 12] we must summarize the results obtained before for perovskites and layered structures.

MAGNETIC DILUTION IN LaMO$_3$-LaAlO$_3$ SYSTEMS

First, let us review the results for perovskites studied as solid solutions LaMO$_3$-LaAlO$_3$ (LaM$_x$Al$_{1-x}$O$_3$) and compare them between themselves along the series of 3d-elements and with solid solutions based on lanthanum gallate (LaM$_x$Ga$_{1-x}$O$_3$).

For the LaTi$_x$Al$_{1-x}$O$_3$ system In the $1/\chi_{Ti}$ vs T plot the deviations from Curie-Weiss law are observed typical for the contribution of temperature independent paramagnetism (TIP), The TIP contribution to paramagnetic compound of magnetic susceptibility decreasing as the titanium concentration increases. Such a behavior cannot be associated with van Vleck TIP or with Pauli paramagnetism (free electrons) only. The matter is that at the infinite dilution and small titanium concentration van Vleck paramagnetism must prevail (single titanium atoms) and at high concentrations of titanium in the solid solutions at the expense of an increase in the number of conductivity electrons (t_{2g}^1) Pauli type of TIP must dominate.

The effective magnetic moment at the infinite dilution, μ_{eff} ($x \to 0$) = 0.86 µB, is substantially lower than the spin only value ($\mu_{s.o.}$(Ti(III)) = 1.73 µB). This fact cannot be explained only by the contribution of spin orbit coupling or by the formation of exchange bonded clusters of titanium atoms. Here we seem to have a substantial (up to 75%) quantity of Ti(IV).

All in all the character of concentration dependence of paramagnetic component on Ti concentration in the solid solution is typical for the dilution of antiferromagnets (Figure 1) and is determined by the formation of clusters of titanium atoms with antiferromagnetic exchange.

For the LaV$_x$Al$_{1-x}$O$_3$ system a break is observed in the temperature dependence of the inverse paramagnetic component of magnetic susceptibility in the region of 240-280 K, Curie-Weiss law obeyed in both parts – low and high temperature (Figure 2).

Electronic Structure of Solid Oxides Doped ...

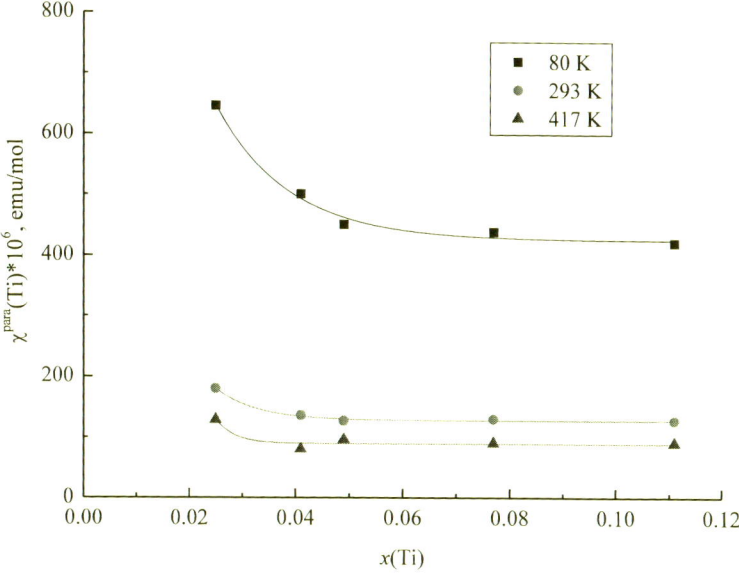

Figure 1. Plots of paramagnetic component of magnetic susceptibility vs Ti content in the $LaTi_xAl_{1-x}O_3$ solid solutions.

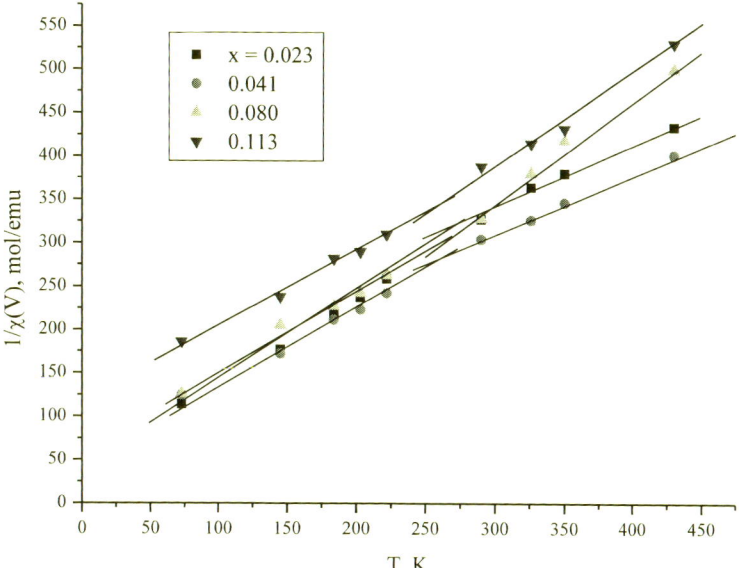

Figure 2. Plots of inverse paramagnetic component of magnetic susceptibility vs temperature for the $LaTi_xAl_{1-x}O_3$ solid solutions.

The concentration dependence of paramagnetic component is also typical for the dilution of antiferromagnets as for LaTi$_x$Al$_{1-x}$O$_3$ (Figure 1). The effective magnetic moments at the infinite dilution μ_{eff} ($x \to 0$) = 2.80 µB are close to the spin only values of $\mu_{S.O.}$(V(III)) = 2.73 µB.

Magnetic dilution in the LaCrO$_3$ – LaAlO$_3$ system is to a large extent similar to the systems containing Ti and V. Concentration dependence of paramagnetic component is also typical for the dilution of antiferromagnets (Figure 4). The values of χ_{Cr} decrease the most significantly in the region of low concentrations of Cr ($x < 0.05$), during further increase in x the changes in χ_{Cr} are less impressive. The extrapolation of isotherms to the infinite dilution results in a constant μ_{eff} ($x \to 0$) = 4.00 µB, which is close to $\mu_{S.O.}$(Cr(III)) = 3.83 µB.

There is also two sites in the plots of 1/χ_{Cr} vs T (as in the LaV$_x$Al$_{1-x}$O$_3$ system) where 1/χ_{Cr} undergoes a jump wise change in the very region of $T \sim 270 - 290$ K (Figure 5) with negative Weiss constant for both sites and for all the concentrations.

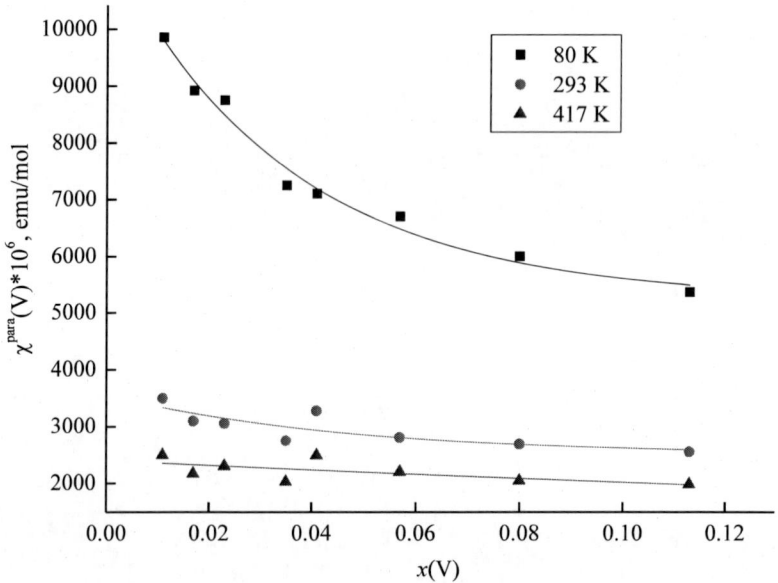

Figure 3. Plots of paramagnetic component of magnetic susceptibility vs V content in the LaV$_x$Al$_{1-x}$O$_3$ solid solutions.

Electronic Structure of Solid Oxides Doped ... 71

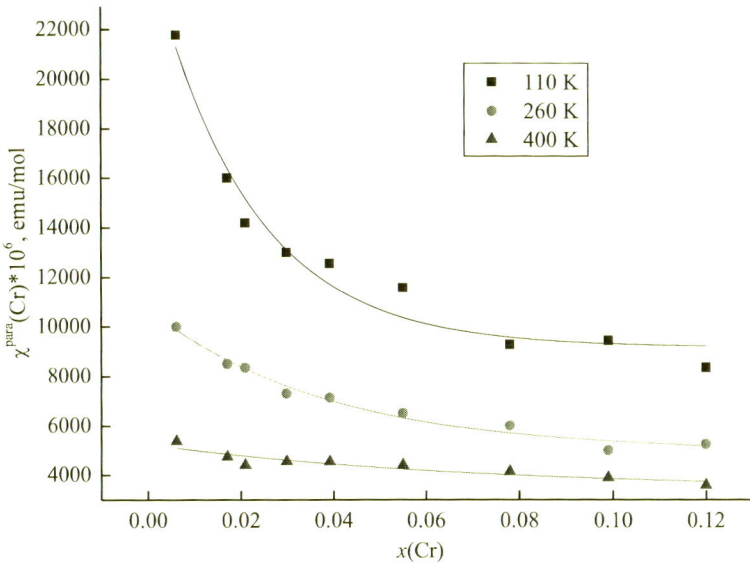

Figure 4. Plots of paramagnetic component of magnetic susceptibility vs Cr content in the $LaCr_xAl_{1-x}O_3$ solid solutions/.

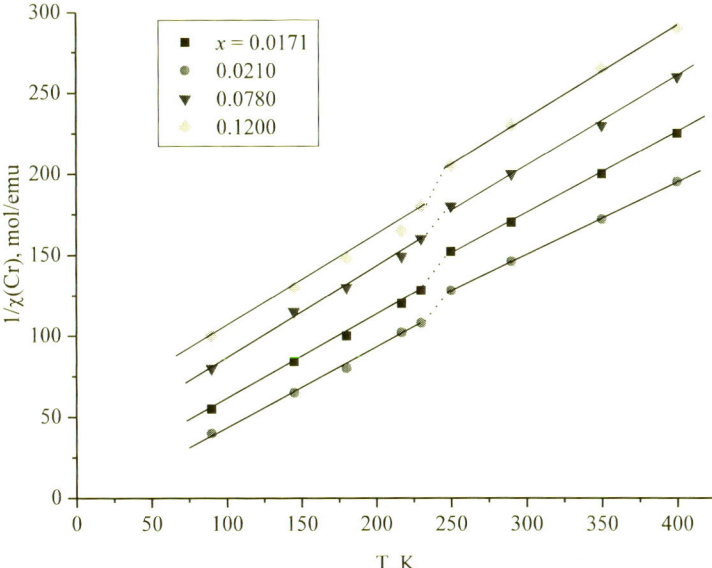

Figure 5. Plots of inverse paramagnetic component of magnetic susceptibility vs temperature for the $LaCr_xAl_{1-x}O_3$ solid solutions/.

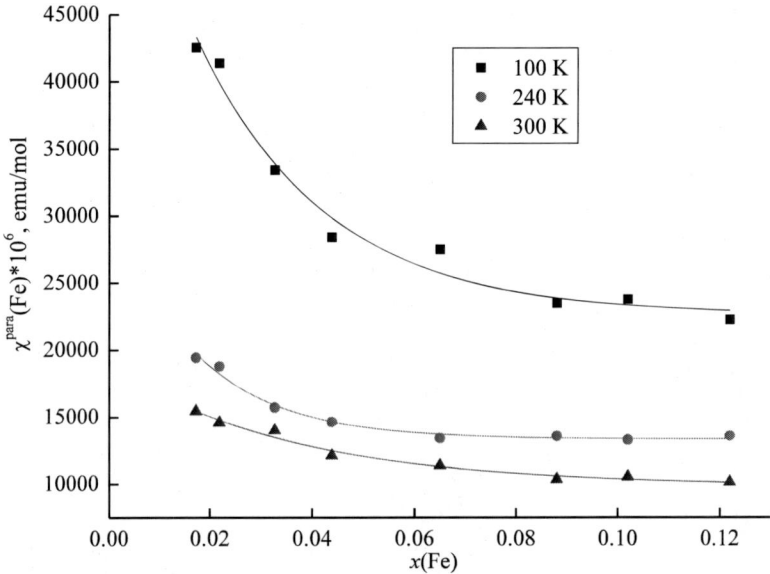

Figure 6. Plots of paramagnetic component of magnetic susceptibility vs Fe content in the $LaFe_xAl_{1-x}O_3$ solid solutions.

The constancy of the temperature of the break in the temperature dependencies of $1/\chi_M$ seems to be associated with the process of localization ↔ delocalization of electrons. The possibility of such transfers is shown by Goodenough in [13-15], where he studied the superexchange at 180° in perovskites. The character and energy of exchange interactions depend on mutual orientation of metal and ligand (oxygen) orbitals and on the occupation of d-orbitals with electrons [4].

All in all magnetic dilution of chromium containing system results in the conclusion about the presence of Cr(III) clusters in the solid solutions with antiferromagnetic exchange.

The dependence of χ_{Fe} for the $LaFe_xAl_{1-x}O_3$ system as well as for the solid solutions containing Co(III) and Ni(III) and for the above mentioned systems is also typical for the dilution of antiferromagnets. Here we observe a rather sharp decrease in χ_{Fe} in the region of small iron concentrations ($x = 0.01 - 0.04$), which points to the formation of antiferromagnetic clusters. Further increase in x favors the formation of larger clusters and interactions between them, which seems to make a smaller impact on magnetic susceptibility.

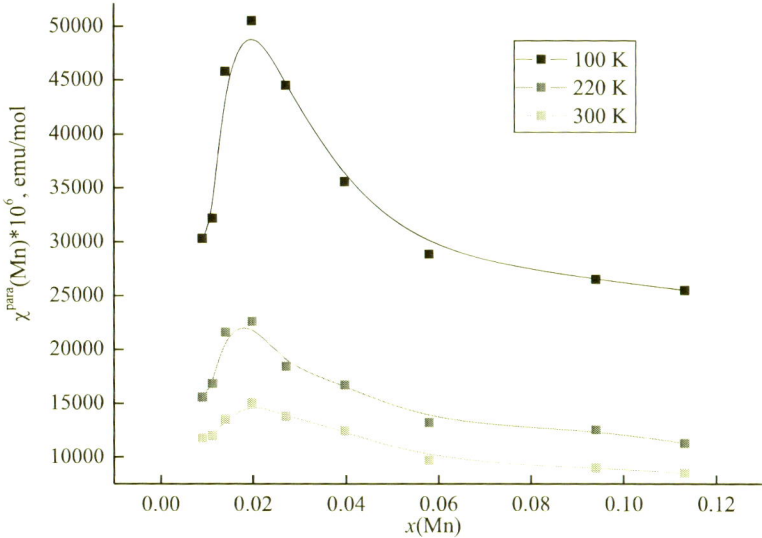

Figure 7. Plots of paramagnetic component of magnetic susceptibility vs Mn content in the LaMn$_x$Al$_{1-x}$O$_3$ solid solutions/.

The extrapolation of magnetic susceptibility isotherms to the infinite dilution gives μ_{eff} ($x \to 0$) = 6.00 µB, which is close to $\mu_{S.O.}$(Fe(III)) = 5.92 µB.

In the plots of 1/χ_{Fe} vs T there is the typical break, but the temperature of the break depends on the iron concentration. Weiss constants change from negative to positive as x increases. Thus for all the solid solutions with $x < 0.04$ $\theta < 0$, and at $x > 0.04$ $\theta > 0$ in the region of high temperatures. In other words we seem to deal with a competition between antiferro- and ferromagnetic types of interactions within the clusters of iron atoms. However, taking into account the exchange channel theory [4], we see that the exchange between e_g electrons must be strongly antiferromagnetic owing to a great overlapping, ferromagnetic superexchange is the result of a crossing superexchange or electron correlation of the type $d_{x^2-y^2} \mid p_x \perp p_z \mid d_{z^2}$ or $d_{xy} \mid p_x \perp p_z \mid d_{z^2}$.

Ferromagnetic superexchange becomes stronger when we have some distortion of the octahedra, which is possible as the concentration of large Fe(III) increases. All this brings about a small absolute value of exchange parameter J for iron in $LaAlO_3$ ($J = -9$ cm^{-1}) in comparison with Cr ($J = -22$ cm^{-1}).

The $LaMn_xAl_{1-x}O_3$ is substantially different from those considered above. In the $\chi_{Mn} - x$ plots there is a well-defined maximum in the region of $x \sim 0.02$ (Figure 7).

Extrapolation of the isotherms of magnetic susceptibility results in μ_{eff} ($x \to 0$) = 4.80 µB, close to $\mu_{S.O.}(Mn(III))$ = 4.73 µB. An increase in the magnetic susceptibility in the region of $x = 0 - 0.02$ cannot be associated with ferromagnetic exchange between Mn(III). According to the exchange channel model J must be negative. The following model was used for the description of such a behavior, based in particular on the data of ESR spectra, where at low concentrations the signals from Mn(II) and Mn(IV) were found. From thermodynamic data we see that the standard formation enthalpies of MnO, Mn_2O_3, and MnO_2 do not differ significantly [16]. Moreover, the exchange Mn(II)-O-Mn(IV) must be ferromagnetic. Thus it was assumed that at low concentrations Mn(III) disproportionates in the dimer clusters to Mn(II) and Mn(IV), which results in ferromagnetic component in the exchange and in an increase in the susceptibility.

This assumption is indirectly confirmed by the values of Weiss constants, which at low concentration (up to $x = 0.02$) are positive, and become negative as x increases. At rather high concentrations ($x > 0.06$) we can observe a small deviation of $1/\chi_{Mn} - T$ plots from linearity, which points to a TIP, which increases as x increases. This allows us to ascribe it to Pauli paramagnetism.

The same result we obtained for nickel containing perovskite, which is supported by the fact that pure $LaNiO_3$ is a metal conductor. At $x > 0.04$ paramagnetic component decreases monotonously, which points to a long order antiferromagnetic exchange typical for pure $LaMnO_3$.

The dependences of paramagnetic component of magnetic susceptibility on transition metal concentration are typical for the dilution of antiferromagnets – a decrease in χ_M as x increases (Figures 1-6), but for manganese containing solid solutions [17] here we meet the situation when manganese (III) disproportionates to Mn(IV) and Mn(II), which results in ferromagnetic exchange between them and an increase in χ_M given $0.01 > x > 0.05$. The disproportionation is confirmed by ESR spectra. Thus, for all the elements under study we appear to have single atoms at zero concentration, even of Mn(III). Another problem showed itself in the electron structure of Co and Ni containing solid solutions: both elements, especially Ni(III) [13], were expected to have low spin configuration. The calculation of the temperature dependence of the effective magnetic moment at the infinite dilution for Co and Ni showed that we deal with spin equilibrium [18] and a comparatively small energy gap between high spin and low spin states.

One of the most important results is the distribution of paramagnetic atoms over the diamagnetic matrix $LaAlO_3$. We found that first, the fraction of dimer clusters in the diluted solutions exceeds their fraction calculated for the statistical distribution by a factor of about 2 (Figure 8) and second, as is seen from Figure 8 the distribution of $3d$-element atoms does not depend on their nature.

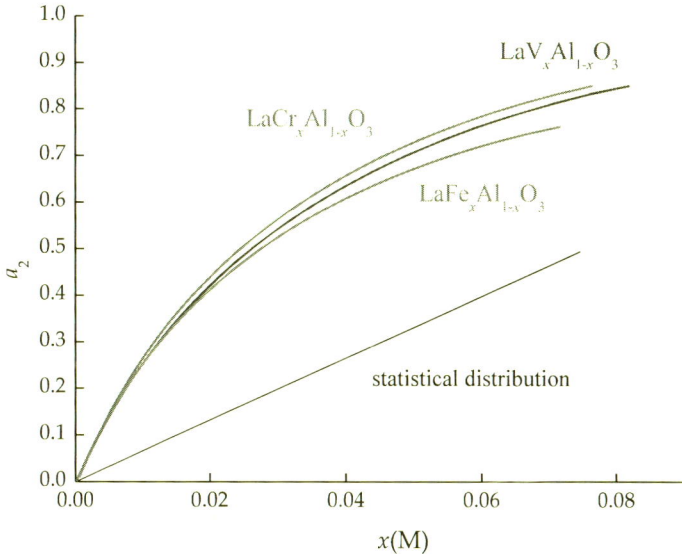

Figure 8. The fractions of dimer clusters in the $LaM_xAl_{1-x}O_3$ solid solutions.

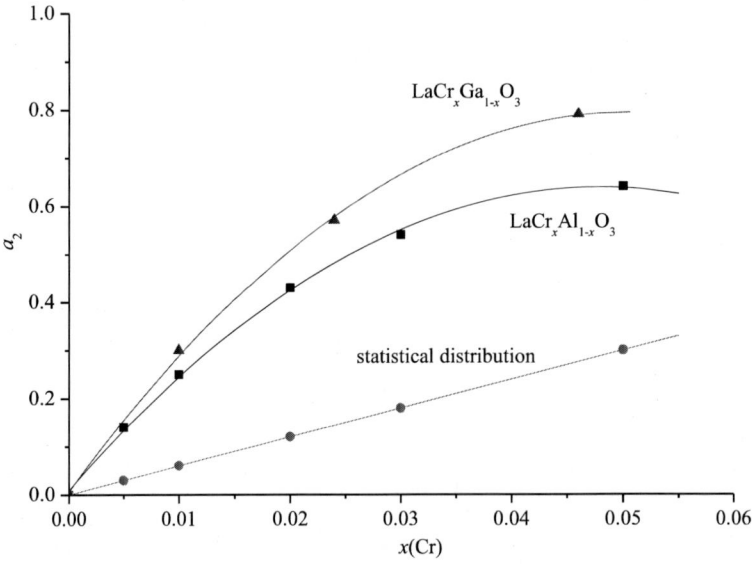

Figure 9. The fractions of dimer clusters in the LaCr$_x$Al$_{1-x}$O$_3$ and LaCr$_x$Ga$_{1-x}$O$_3$ solid solutions.

An estimation of the interchange energy by [5] gave about 12 ± 1 kJ/mol. Furthermore, this interchange energy does depend on the nature of diamagnetic elements in the matrix, however it strongly depend on the composition of diamagnetic matrix (Figure 9). This energy in the solid solutions of chromium in LaGaO$_3$ gives 15 ± 1 kJ mol. We must say, that such a distribution of paramagnetic atoms was confirmed in our ab initio calculations taking solid solutions containing iron(III) as an example [19]. It is interesting that the estimation of the exchange parameter for chromium(III) between two systems based on lanthanum aluminate and lanthanum gallate results in $J = -18$ cm^{-1} for the former and $J = -12$ cm^{-1} for the latter. This means that we observe a decrease in the antiferromagnetic exchange upon passing from aluminate to gallate and a simultaneous increase in the interatomic interactions of nonmagnetic character. First of all we emphasize, that the interactions responsible for the distribution of paramagnetic atoms in a diamagnetic matrix have nothing in common with magnetic exchange – the energies are incompatible. So, let us discuss the reasons for both effects.

Lanthanum aluminate crystallizes in the rhombohedral syngony, whereas $LaGaO_3$ as well as the most part of perovskites containing lanthanum and larger *d*-elements have rhombic structure. Orthorhombic structure implies the changes in the M-O-M angles in the perovskite layer. The superexchange between chromium atoms in perovskite structure occurs via $d_{xy} \parallel p_y \parallel d_{xy}$ and $d_{xz} \parallel p_z \parallel d_{xz}$ exchange channels, i.e., via p_π-d_π overlapping of chromium orbitals with oxygen orbitals [4]. There is a ferromagnetic superexchange channel - $d_{xy} \parallel p_y \perp p_z \parallel d_{yz}$. A decrease in the exchange angle in lanthanum gallate results in a decrease in the antiferromagnetic exchange, whereas ferromagnetic interaction independent on the orbital overlapping remains unchanged.

As for the distribution of paramagnetic atoms in the structure, there are two factors to be considered – the atom sizes and the ionicity of their bond with oxygen. The radius of gallium atom is close to the radius of chromium(III) [20]. Hence no distortions of the nearest surrounding in lanthanum gallate upon introduction of chromium could be expected. Such local distortions must occur upon substitution of chromium for small aluminum (R=0.53 A). This circumstance must prevent 3*d*-metal aggregation due to spatial difficulties. On the other hand, an increased ionicity of competing bonds diamagnetic element (Ga) – oxygen must result in an increased covalence of Cr-O bond and an increase in the clustering of paramagnetic element.

In other words from close inspection of magnetic dilution in perovskite matrices we could derive several regularities, concerning the elements in Periodic table.

1. The distribution of 3d-elements over the same diamagnetic matrix does not depend on their nature.
2. The distribution of paramagnetic elements does depend on the composition of a diamagnetic matrix being associated with bond ionicity of diamagnetic elements with oxygen.
3. The interchange energy responsible for the distribution of paramagnetic elements over a diamagnetic matrix seems to be

associated with the interaction of filled electron orbitals, and its energy is counted no less than 10 and more kJ/mol, whereas magnetic exchange is due to the interaction of unpaired electrons, its energy never exceeding 1 kJ/mol.

4. Magnetic superexchange interactions obey the model of exchange channels, J depending on the state of paramagnetic atoms and also on the ionicity of metal-oxygen bond.

MAGNETIC DILUTION IN PEROVSKITE-LIKE LAYERED STRUCTURE

Complex oxides with layered K_2NiF_4 structure besides their promising properties as materials for electronics offer new possibilities for the development of magnetic dilution method and searching for the regularities in the changes in the electronic structure of solids. The tetragonal structure of K_2NiF_4 type (Figure 10) suggests that the AA'BO$_4$ complex oxides must be 2D-magnetics.

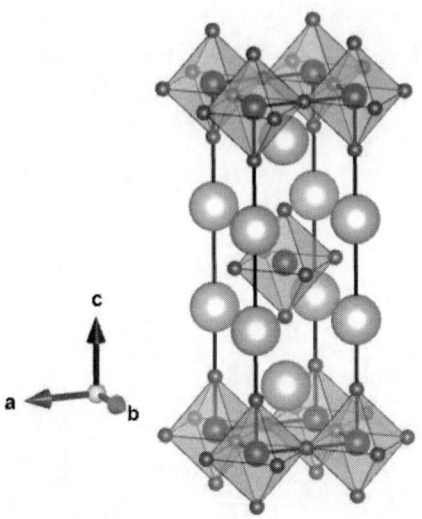

Figure 10. Structure of K_2NiF_4 type.

But for the study of exchange interactions between paramagnetic atoms and their electronic state this seems to simplify the task, since we have only the exchange interactions in the *xy* plane formed by oxygen octahedra. Moreover, there is a possibility to vary not only the B diamagnetic atoms as in perovskite, but also both A and A' and study the dependences of the chemical structure on the nature of the so called large atoms, for example changing Ls and Sr for Y and Ca.

We studied solid solutions with the general formulae $LaSrM_xAl_{1-x}O_4$ and $YCaM_xAl_{1-x}O_4$, where M – Cr, Mn, Fe, Co, and Ni. However, magnetic dilution in these oxide systems appeared to present a number of specific features difficult to interpret. First of all, only for chromium substituted systems the regular dependence of paramagnetic susceptibility on the concentration was observed. If for Mn containing systems we observe the same trend as in perovskite – disproportionation of Mn and ferromagnetic exchange in Mn(II)-Mn(IV) dimers. The magnetic behavior of Fe, Co, and Ni containing solid solutions required special consideration.

So let us consider the chromium containing solid solutions. Cr(III) is the most stable state for chromium atoms, moreover, its ground state is $^4A_{2g}$. Non-degenerate ground state allows Heisenberg-Dirac-van Vleck model to be used quite accurately, since this theory is limited by using *A* and *E* ground states (Figure 11). Using experimental data for both layered oxides containing Cr(III) and also the $Sr_2Mn_xTi_{1-x}O_4$ system, where Mn(IV) has the same electron configuration d^3, we found the following. Exchange parameter *J* decreases in absolute value from -18 for LaSr-system to -13 cm^{-1} for YCa-system, which can be ascribed both to a local decrease in the Cr-O-Cr angle in the structure of smaller size and to an increase in the ionicity of Cr-O bond. All these factors decrease the overlapping of *d*-orbitals with *p*-orbitals of oxygen, thus decreasing the antiferromagnetic exchange. However for $Sr_2Mn_xTi_{1-x}O_4$ the antiferromagnetic exchange increases to -30 cm^{-1}. This points to the fact the an increased covalence of Mn(IV)-O bond becomes the crucial factor For the exchange. Moreover, the distribution of paramagnetic atoms in the layered matrix also shows an increased clustering as the M-O bond becomes more covalent (Figure 12) [21].

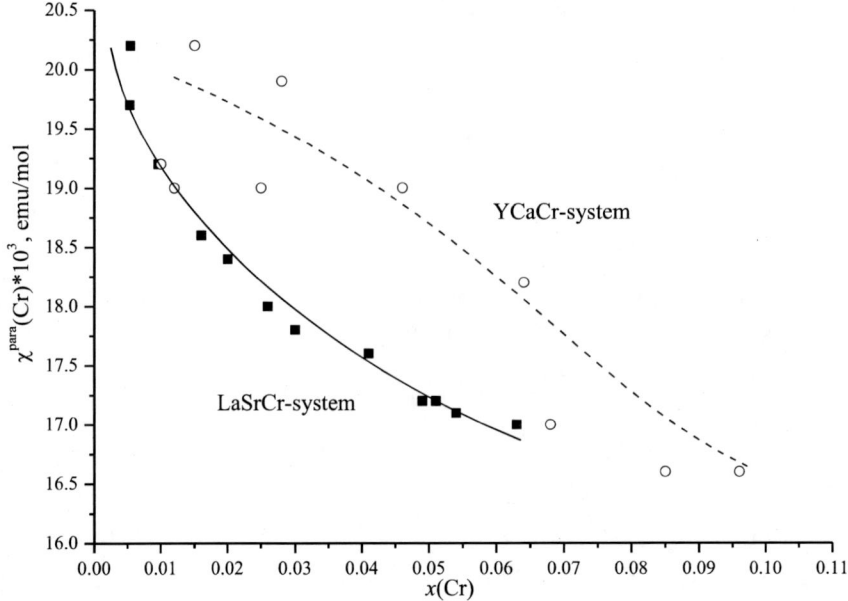

Figure 11. Plots of paramagnetic component of magnetic susceptibility vs Cr content in the LaSrCr$_x$Al$_{1-x}$O$_4$ and YCaCr$_x$Al$_{1-x}$O$_4$ solid solutions.

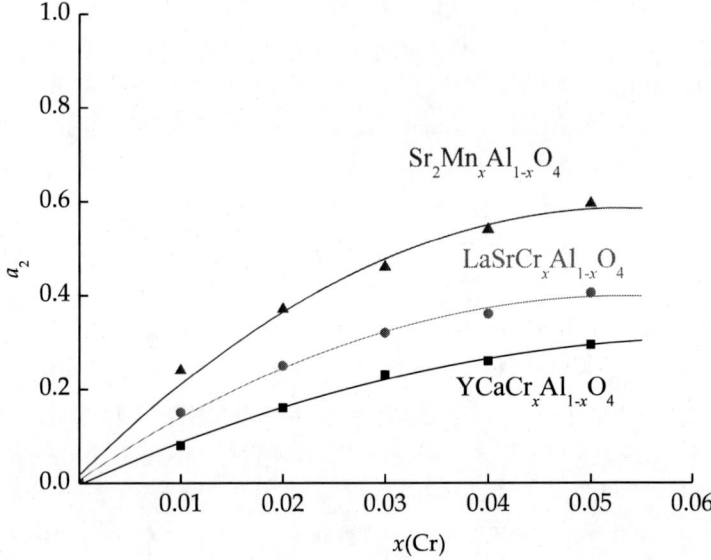

Figure 12. Plots of dimer fractions vs concentration of chromium or magnesium for YCaCr$_x$Al$_{1-x}$O$_4$, LaSrCr$_x$Al$_{1-x}$O$_4$ and Sr$_2$Mn$_x$Ti$_{1-x}$O$_4$ solid solutions.

Electronic Structure of Solid Oxides Doped ... 81

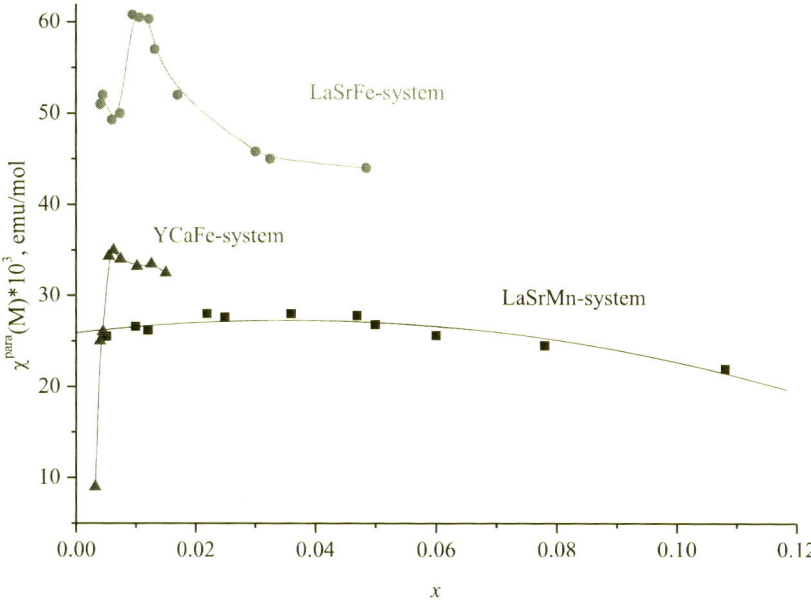

Figure 13. Plots of paramagnetic component of magnetic susceptibility vs Fe or Mn content in the $LaSrFe_xAl_{1-x}O_4$, $YCaFe_xAl_{1-x}O_4$ and $LaSrMn_xAl_{1-x}O_4$ solid solutions.

Iron and manganese seem to be too large for the $YCaAlO_4$ matrix to carry out a comparison of the properties of solid solutions. In the $LaSrAlO_4$ matrix manganese disproportionates like in the perovskite, whereas Fe(III) demonstrates complex magnetic dilution with alternating antiferro-ferro-antiferromagnetic type of exchange depending on the changes in the octahedra distortions (Figure 13) [22]. In the $YCaAlO_4$ matrix iron is oxidized to Fe(IV), which was proved by Moessbauer studies.

Leaving aside iron containing layered oxides, we must pay attention to cobalt and nickel containing systems. The problem with these elements is that they both can have different spin states. Nickel can be low spin, high spin and spin equilibrium. As for Co(III) it can also be in the intermediate spin state ($S = 1$) when the distortion of the octahedra is substantial.

We begin with nickel containing systems. The isotherms of χ_{Ni} (Figure 14) show an increase in paramagnetic component with a maximum at $x \sim 0.07$ for LaSr-system and at $x \sim 0.02 - 0.03$ for YCa-system, those latter lying lower than the former.

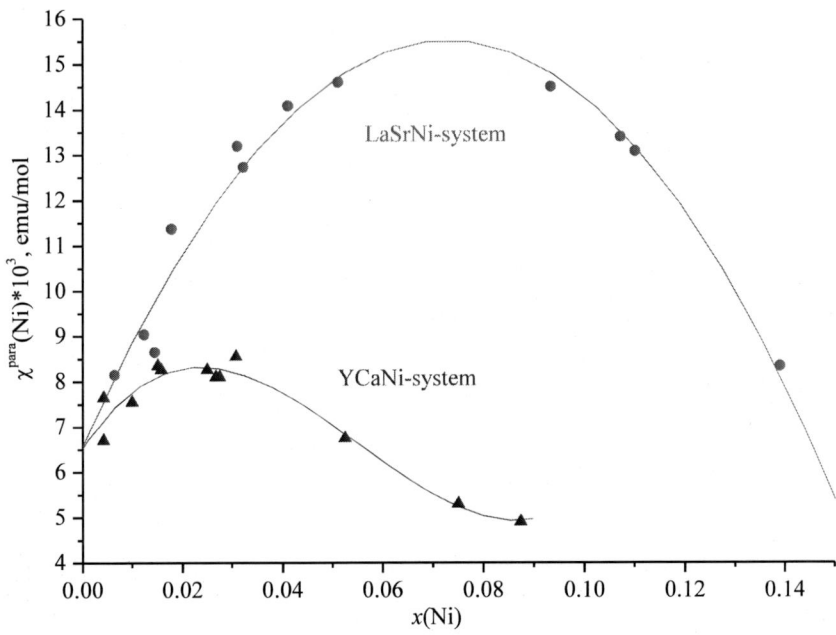

Figure 14. Plots of paramagnetic component of magnetic susceptibility vs metal content in the LaSrNi$_x$Al$_{1-x}$O$_4$ and YCaNi$_x$Al$_{1-x}$O$_4$ solid solutions at 80 K.

Extrapolating the isotherms to the infinite dilution we obtain the effective magnetic moments depending on temperature and essentially lower than could be expected for high spin Ni(III). An attempt to calculate the μ_{eff} within the assumption of low spin-high spin equilibrium failed – we could not obtain a single value of the energy of transition as we were able to do in the case of perovskite structure. There was one possible assumption to be made: in the K$_2$NiF$_4$ – type structure with two various atoms in the sites with coordination number 6 the octahedral sites with various surrounding by heavy metals exist taking into account the fact that heavy atoms are distributed statistically. This results in the existence of nonequivalent crystal fields with different distortions of the octahedra and different ionicity of Ni-O bond. Therefore for both systems we must consider the possibility of various spin states and various interactions in the clusters.

In LaSrNi$_x$Al$_{1-x}$O$_4$ the extrapolation of isotherms of paramagnetic component of magnetic susceptibility gives μ_{eff} = 1.78 μB independent on temperature, which corresponds to Ni(III) low spin ($t_{2g}^6 e_g^1$, ^2E_g). But the magnetic susceptibility in the region of maximum is too large to be ascribed to any ferromagnetic interactions in small clusters of low spin nickel atoms ($\mu_{eff} \sim 3.0$ µB increasing with temperature). This suggests, that as concentration increases, a fraction of high spin Ni(III) appears in the structure, which results in ferromagnetic exchange between Ni(III)$_{ls}$ and Ni(III)$_{hs}$. Crystallographic data confirm this suggestion since the octahedra tetragonal distortion in LaSrAlO$_4$ is 1.069 [23], whereas in LaSrNiO$_4$ it is only 1.030 [24]. By the way, in YCaAlO$_4$ it is even greater – 1.094 [25]. Comparatively large distortions of the octahedra favor the existence of Ni(III) in the low spin state, so we obtain such a state in almost pure LaSrAlO$_4$. An increase in Ni concentration results in a decrease in the distortion and the appearance of high-spin Ni(III). Then we could try to carry out the calculations of exchange parameters within the suggestion that Ni(III) monomers are low spin, the dimers are ½-1/2, ½-3/2 and 3/2-3/2.

Here another problem arises. When dealing with triplet states (Ni(III), $^4T_{2g}$) we must consider both spin-orbit splitting and exchange splitting [4]. We cannot afford such calculations due to the limits of the accuracy of our extrapolations. However we could use a temperature dependent g-factor for a triplet term as was suggested in [26]. We calculated g-factor for low spin state of Ni(III) taking $\mu_{S.O.} = 1.78$ µB obtained at the infinite dilution as $2\mu_{exp}/\mu_{S.O.}$. And estimated the spin orbit coupling constant as -220 cm^{-1} from the formula

$$g = 2 - \frac{\lambda}{10 Dq} \text{ with } 10Dq = 17000 \text{ cm}^{-1} \text{ [27]}$$

Since the single electron spin orbit coupling constant ξ must not depend on the ground state we could calculate the g values for the high spin state at various temperatures and use them in our calculations. The parameters of antiferromagnetic exchange (3/2-3/2, since in the layered structure d_z^2 orbitals do not overlap, therefore such dimers behave as monomers) obtained from our calculations are -60 for LaSr- and -40 for

YCa-system. The parameter of ferromagnetic exchange The general trend is the same as for chromium containing systems and in agreement with exchange cannel model, since the greatest contribution into magnetic exchange is due to $d_{x^2-y^2} \| p_x \| d_{x^2-y^2}$ channel with $J << 0$ [4]. Ferromagnetic exchange parameter remains the same for both systems $J_{1/2\text{-}3/2} = +10$ cm^{-1}.

For cobalt containing layered oxides we must take into account the possibility of cobalt atoms being in low spin state ($S = 0$, $^1A_{1g}$), high spin ($S = 2$, $^4T_{2g}$), and intermediate spin state ($S = 1$, $^3T_{1g}$), the latter being realized in the structures with substantial tetragonal distortion of oxygen octahedra [28].

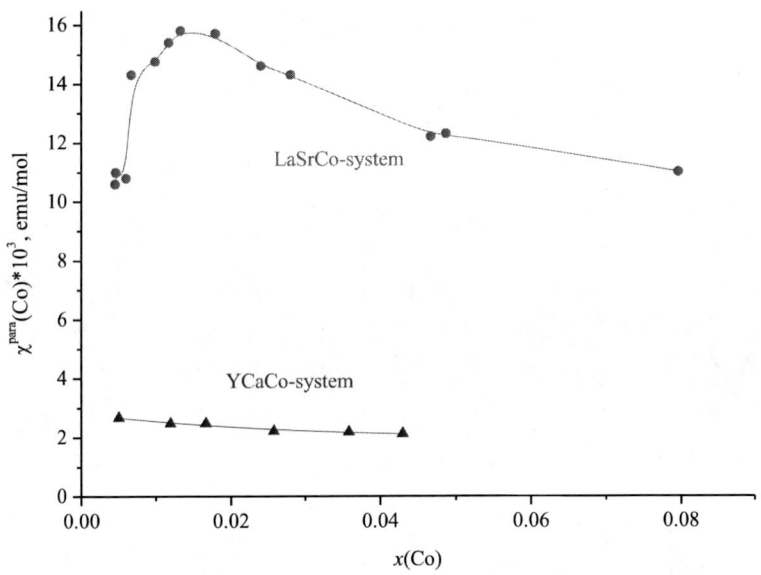

Figure 15. Plots of paramagnetic component of magnetic susceptibility vs metal content in the LaSrCo$_x$Al$_{1-x}$O$_4$ and YCaCo$_x$Al$_{1-x}$O$_4$ solid solutions at 80 K.

The calculation of the paramagnetic susceptibilities for the whole range of solid solutions LaSrCo$_x$Al$_{1-x}$O$_4$ and YCaCo$_x$Al$_{1-x}$O$_4$ (Figure 15) using the formulae [18] showed that in contrast to Ni containing systems we do have the spin equilibrium $^3T_{1g} \leftrightarrow {}^1A_{1g}$, the fraction of paramagnetic cobalt increasing with temperature since the ground state is $^3T_{1g}$ and the energy difference being ~ 110 cm^{-1} for LaSr- and 75 cm^{-1} for YCa-system. The exchange parameter for magnetic interactions between paramagnetic cobalt atoms is ~ -20 cm^{-1} LaSr-system [29] and decreases for YCa-system. This is also in agreement with exchange channel model, since in the exchange between paramagnetic cobalt atoms occurs via d_{xy} orbitals, taking into account that unpaired electrons in cobalt with the intermediate spin are in d_{xy} and d_{z^2} orbitals (lower in an elongated octahedron).

As a consequence we could derive general trends for the changes in the exchange parameters and in the interatomic interactions based on the ionicity of the competing bonds in AA'BO$_4$ layered perovskite like oxides (Table 1). In the matrices with higher ionicity of A-O bond and therefore higher covalence of B-O bond the exchange parameter appears to be greater.

Antiferromagnetic exchange parameter increases in absolute value in the sequence YCaAlO$_4$ < LaSrAlO$_4$ < Sr$_2$TiO$_4$ (J for Mn(IV) in Sr$_2$TiO$_4$ is -30 cm^{-1}). The clustering of paramagnetic atoms reflecting the interchange energy also changes in the same sequence.

Table 1. Exchange parameters for various transition metals in LaSrAlO$_4$ and YCaAlO$_4$ matrices

Diamagnetic matrix	J, cm^{-1}		
	Cr(III), d^3	Co(III), d^6	Ni(III), d^7
LaSrAlO$_4$	-18	-20	-60
YCaAlO$_4$	-13	-13	-40

OXIDES WITH SPINEL STRUCTURE

Complex oxides with spinel structure never ceased to attract attention of the researchers [30-32]. But two main problems of electronic structure of such oxides – the distribution of elements over two nonequivalent sites of spinels and the interatomic interactions often remain beyond consideration. The distribution of the elements depends on the energy of preference to octahedral sites [33]. The magnetic dilution method also appears to be instrumental for solving these problems.

We shall discuss two spinel systems based on classic $MgAl_2O_4$ and Mg_2GeO_4. The possibility of paramagnetic atoms distributing over octahedral and tetrahedral sites influences the magnetic properties, exchange interactions, temperatures of magnetic ordering of such materials. For multicomponent spinels the examination of the results of X-ray diffraction methods not always gives unambiguous information about the degree of spinel inversion. This results in a divergence in published data on the energy of preference of elements to octahedral sites.

The distribution of paramagnetic elements over two nonequivalent sites may be derived from the data of magnetic properties of diluted solid solutions. As for solid solutions based on Mg_2GeO_4, this oxide exist in two polymorphic modifications – spinel and olivine. In the olivine structure M(II) atoms occupy only octahedral sites. The studies of the solid solutions $M_xMg_{2-x}GeO_4$ (M – Fe, Co, Ni) with both structures obtained from one another (olivine transforms into spinel under high pressure) showed that magnetic characteristics coincide with the accuracy of 2% over the whole interval of temperatures and concentrations. This proved that in the germanate spinels d-elements are in octahedral sites. This could be expected, since it is common knowledge that germanium like silica are usually in tetrahedra of oxygen atoms.

$MgAl_2O_4$ is known to be the so called normal spinel. The magnetic dilution method allows the distribution of d-elements over octahedral and tetrahedral sites to be determined. The method is based on the following principles. First, at the infinite dilution, if we have only single atoms of a paramagnetic, their susceptibilities are summed up with a fraction of each kind of atoms in both sites; therefore, the effective magnetic moments may be represented as:

$$\mu_{eff}^2 = a\mu_O^2 + (1-a)\mu_T^2, \tag{8}$$

where μ_O and μ_T are the effective magnetic moments of a particular atom in octa- and tetrahedral sites respectively and a is the fraction of paramagnetic atoms in the octahedral sites. The problem is that to calculate the effective magnetic moments we must know the values of the spin orbit coupling constants, which in solids may be lower than the values in the complex compounds due to the spin orbit coupling being somewhat "frozen" [4]. This is supported by our own numerous data (Table 2). The second problem is that d-elements have different ground states in octahedra and tetrahedra. But fortunately, if in one surrounding the ground state (the case of weak field) is A or E, for which μ_{eff} does not depend on temperature, the other is necessarily a triplet state. The only exclusion is d^5 electron configuration.

The Table shows that in each case we have one μ_{eff}, which does not depend on temperature, but its value for A and E ground terms is determined by formula:

$$\mu_{eff} = \mu_{S.O.}\left(1 - \frac{\alpha\lambda}{10Dq}\right)$$

where $\mu_{S.O.} = [4S(S+1)]^{1/2}$ is the spin only effective magnetic moment, $\alpha = 4$ for A state and 2 for E state, λ – spin orbit coupling constant, $10Dq$ – crystal field splitting [5]. The crystal field splitting for all the bivalent elements of iron group in the oxygen surrounding may be taken as 10 000 cm^{-1}, which is supported by spectral studies of spinels [33]. Now to estimate the values of λ we suggest the following procedure. The effective magnetic moment for triplet ground states is a function of kT/λ [5]. Both plots of μ_{eff} cross each other at a certain point together with the experimental plot of μ_{eff} vs kT/λ. This allows the estimation of spin orbit constants for the elements under study.

Table 2. Ground states of various electron configurations in octahedral and tetrahedral surrounding

Electron conFigure		d^1	d^2	d^3	d^4	d^5	d^6	d^7	d^8	d^9
Ground state	O_h	$^2T_{2g}$	$^3T_{1g}$	$^4A_{2g}$	^5E_g	$^6A_{1g}$	$^5T_{2g}$	$^4T_{1g}$	$^3A_{2g}$	^2E_g
	T_d	^2E	$^3A_2$	$^4T_1$	$^5T_2$	$^6A_1$	^5E	$^4A_2$	$^3T_1$	$^2T_2$

Table 3. Theoretical and calculated spin-orbit coupling for some 3d-elements in spinels

λ, cm^{-1}	Co(II)	Ni(II)	Cu(II)
Published data for coordination compounds [3]	-172	-315	-830
Our estimation for Mg$_{1-x}$M$_x$Al$_2$O$_4$	-120	-80	-450

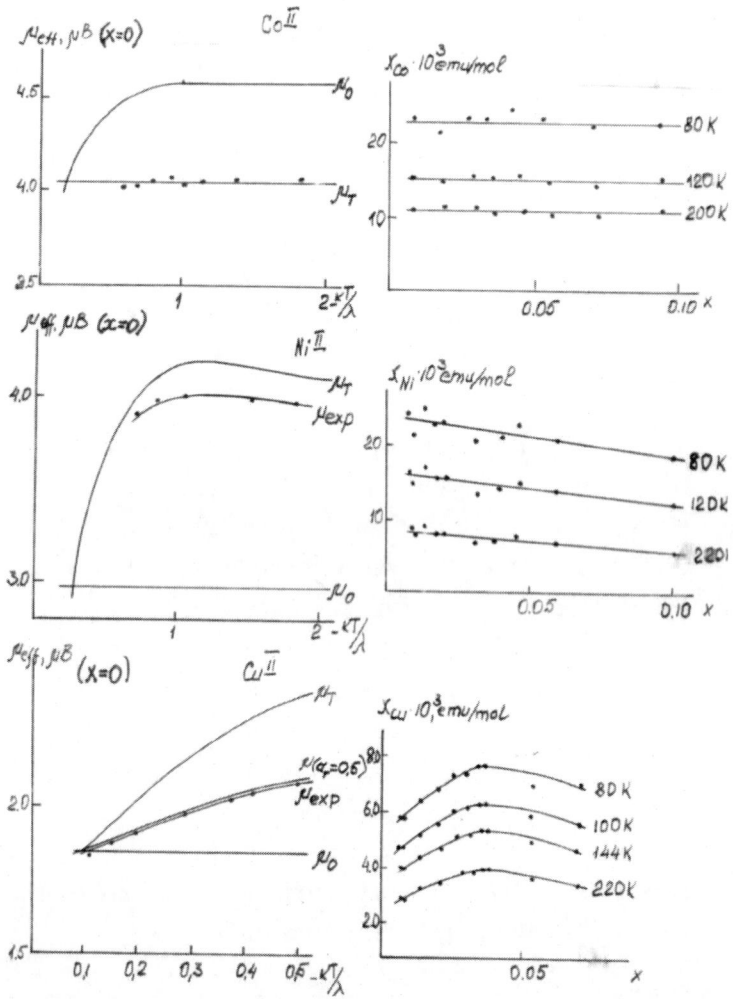

Figure 16. Effective magnetic moment at the infinite dilution for Mg$_{1-x}$M$_x$Al$_2$O$_4$ (M = Co, Ni, Cu) (left) and plots of paramagnetic component of magnetic susceptibility vs metal content in the Mg$_{1-x}$M$_x$Al$_2$O$_4$ solid solutions at various temperatures (right).

As can be seen from Table 3, the spin orbit coupling appears to be substantially less than in complex compounds in complete agreement with the suggestion advanced in [4].

The procedure of estimation of the fraction of paramagnetic atoms in particular sites by formula (8) can be seen in Figure 16. In such a way we found that all cobalt atoms, 85% of nickel atoms and 50% of copper atoms are in tetrahedral sites [34].

This agrees with estimations of the energy of preference for octahedral sites obtained from thermodynamic data by Reznitskii [35]. By the way, all the authors agree that the greater energy of preference for octahedral surrounding is for Cr(III) and for the solid solutions $MgCr_xAl_{1-x}O_4$ at the infinite dilution the effective magnetic moment did not depend on temperature and was equal to 3.77 μB, which means that all the atoms of chromium are in octahedral sites [36].

An essentially more complicated situation with interatomic interactions is observed in the case of germinate spinels [37]. The values of paramagnetic component of magnetic susceptibility and of the effective magnetic moments at the infinite dilution of the $M_xMg_{2-x}GeO_4$ (M – Fe, Co, Ni) solid solutions appears to be greater than is possible for single paramagnetic atoms, which points to the existence of clusters of metal atoms even at the infinite dilution. Such clusters according to the thermodynamic estimations mentioned before [1] must have a large bonding energy. Such a contingency very important for further consideration of interatomic interactions in spinels was to be substantiated. The main conflict consists in the fact that the exchange interactions between, for example, two nickel atoms owing to its two unpaired electrons in e_g orbitals must be strongly antiferromagnetic. We solved the problem in the following way. A qualitative scheme of molecular orbitals in a Ni_2O_{10} cluster does show that *d*-electrons of nickel atoms are paired in the antibonding *s-p* levels. However if we assume that germanium atoms also take part in the formation of clusters and construct a qualitative scheme of molecular orbitals for $Ni_2Ge_3O_{10}$, we can show that the transfer of a part of electron density to the free *d*-orbitals of germanium allows the existence of such a cluster with 6 unpaired electrons. This fact once again shows the role of diamagnetic atoms in the energy and character of interatomic interactions between paramagnetic atoms.

CONCLUSION

In this small chapter we showed the development of the method of magnetic dilution for the studies of the electronic structure of complex oxide systems with isovalent substitution of paramagnetic elements for diamagnetic atoms. These studies paved the way for further investigations of the electronic structure of oxide systems with heterovalent substitution. These systems are now widely used as materials for energy saving technologies, in electronics, etc. Now using the developed method, the obtained regularities of the changes in the properties of complex oxides in the Periodic system of elements we successfully describe a wide series of oxide systems with heterovalent susbstitution used as materials for solid oxide fuel cells, colossal magnetoresistors and so on, which can be found in [1].

REFERENCES

[1] *Electronic structure of materials: challenges and developments.* Chezhina, N. and Korolev, D. Eds. (2019). Pan Stanford Publishing.
[2] Blocker, R. J. and West, F. G. (1969). On magnetic susceptibility in randomly diluted systems. *Phys. Lett.,* 28A: 487-488.
[3] Earnshaw, A. (1968). *Introduction to magnetochemistry.* London & New York: Academic press.
[4] Rakitin, Yu. V. and Kalinnikov, V. T. (1994). *Sovremennaya Magnetokhimia* [in Russian] (*Modern Magnetochemistry*). Sankt-Petersburg: Nauka.
[5] Smirnova, N. A. (1973). *Metody statisticheskoi Termodinamiki v fizicheskoi Khimii* [in Russian] (*Methods of statistical Thermodynamics in Physical Chemistry*). Moscow: Vysshaya shkola.

[6] Kozheurov, V. A. (1975). *Statisticheskaya Termodinamika* [in Russian](*Statistical Thermodynamics*). Moscow: Metallurgiya.
[7] Zhou, W., Ma, C., Cao, M., Gan, Zh. W, Wang, X., Ma, Y., Tan, W., Wang, D., Du, Y. (2017) Large magnetocaloric and magnetoresistance effects in metamagnetic $Sm_{0.55}(Sr_{0.5}Ca_{0.5})_{0.45}MnO_3$ manganite. *Ceram. Int.*, 43(10), 7870-7874.
[8] Jacobson, AJ., (2010). Materials for solid oxide fuel cells. *Chem. Mater.* 22(3), 660-674.
[9] Yaremchenko, A. A., Valente, A. A., Kharton, V. V., Tsipis, E. V., Frade, J. R., Naumovich, E. N., Rocha, J., and Marques F. M. B. (2003) Oxidation of dry methane on the surface of oxygen ion-conducting membranes, *Catalysis Letters*, 91(3–4), 169-174.
[10] Tropin, E., Ananyev, M., Porotnikova, N., Kholimchuk, A, Saher, S., Kurumchin, E., Shepel, D., Antipov, E., Istomin, S., Baunwmeester, H. (2019), Oxygen surface exchange and diffusion in $Pr_{1.75}Sr_{0.25}Ni_{0.75}Co_{0.25}O_{4-\delta}$. *Phys. Chem. Chem. Phys.* 21(9), 4779-4790.
[11] Chezhina N., Korolev, D. (2013) Electron structure of perovskite electron-ionic conductors. *Perovskite. Crystallography, chemistry, and catalytic performance I-st edition*, Ed. J. Zhang, H. Lee New-York: Nova Science Publishers Inc. 199-219.
[12] Chezhina, N., Korolev, D., Zharikova, E. (2014) Influence of the nature and concentration of dia- and paramagnetic doping elements electron structure of doped lanthanum gallate. *Recent advances in gallate research.* A. L. Kinsey Ed., New York. Nova Publishers, 133-200.
[13] Goodenough, J. B. (1971) Coexistence localized and collective d-electrons. *Mat. Res. Bull.* 6, 967-976.
[14] Goodenough, J. B. (1966) Covalency criterion for localized мы collective electrons in oxide with the perovskite structure. *J. Appl. Phys.* 37, 1415-1422.
[15] Goodenough, J. B. (1968) Localized vs collective descriptions of magnetic electrons. *J. Appl. Phys.* 39, 403-407.

[16] Daubert, T. E., Danner, R. P., Sibul, H. M., and Stebbins, C. C., (1994) *Physical and Thermodynamic Properties of Pure Compounds: Data Compilation*, Taylor & Francis, Bristol, PA.

[17] Brach, B. Ya., Dudkin, B. N., Chezhina, N. V. (1979) Magnitnye svoistva tverdykh rastvorov slozhnykh oksidov so strukturoi perovskite, soderzhashchikh 3d-elementy [in Russian] Magnetic properties of solid solutions of complex oxides with perovskite structure containing 3d-elements. *Zh. Neorg. Khim.* 24(8), 2064-2067.

[18] Martin, R. L., White, A. H. (1968) The Nature of the transition between high-spin and low-spin octahedral complexes of the transition metals. *Trans. Met. Chem.* 4, 113-198.

[19] Evarestov, R. A., Bandura, A. V., Sapova, M. D., Korolev, D. A., Chezhina, N. V. (2020) Parameterization of dilute Ising model for iron-containing lanthanum gallate and aluminate solid solutions based on first principle calculations. *Solid State Ionics.* 348, 115283-115291.

[20] Shannon, R. D., Prewitt, C. T. (1969) Effective ionic radii in oxides and fluorides *Acta Cryst.* (B), 25(8), 925-946.

[21] Chezhina, N. V. and Patii, V. P. (2000). State of Atoms and Interatomic Interactions in Complex Perovskite-like Oxides: XIII. Interactions between Mn(IV) Atoms in $Sr_2Mn_xTi_{1-x}O_4$ Solid Solutions. *Rus. J. Gen. Chem.*, 70, 1337-1339.

[22] Chezhina, N. V., Zvereva, I. A., Bobrysheva, N. P. (1992). Sostoyanie atomov zheleza i obmennye vzaimodeistviya v tverdykh rastvorakh $LaSrFe_xAl_{1-x}O_4$ i $YCaFe_xAl_{1-x}O_4$ [In Russian] (State of iron atoms and exchange interactions in the $LaSrFe_xAl_{1-x}O_4$ and $YCaFe_xAl_{1-x}O_4$ solid solutions.) *Zh. Neorg. Khim.* 37(3), 549-554.

[23] Ganguly, P., Rao, C. N. R. (1984) Crystal chemistry and magnetic properties of layered metal oxides possessing the K_2NiF_4 or related structure. *J. Solid State Chem.* 53, 193-216.

[24] Rayom, G., Daire, M. (1977) Contribution a la crystallochimie des composes $SrLnMO_4$ et $CaLnMO_4$ de structure type K_2NiF_4 [in French]. (Contribution to the crystal chemistry of $SrLnMO_4$ and

CaLnMO$_4$ compounds with K$_2$NiF$_4$ type structure). *Revue de chimie mineral*, 14, 11-19.

[25] Choisnet, J., Archaimbault, F., Crespin, M., Chezhina, N., Zvereva, I. (1993) Crystal chemistry of the diluted solid solution YCaCr$_x$Al$_{1-x}$O$_4$ (x ≤ 0.07) *Eur. J. Solid State and Inorg. Chem.* 30(2), 619-628.

[26] Lines, M. E. (1972) Orbital angular momentum in the theory of paramagnetic clusters. *J. Chem. Phys.* 55 (6), 2977-2984.

[27] Demazeau, G., Marty, J. L., Buffat, B., Dance, J. M., Pouchard, M., Dordon, P., Chevalier, B. (1982) Sur la stabilization dans un reseau oxygene du nickel(III) a spin fort en environment octahedrique [in French] (About stabilization of high spin nickel(III) in an oxygen structure in octahedral surrounding) *Mat. Res. Bull.*, 17(1), 37-45.

[28] Buffa, B., Demazeau, G., Pouchard, M., Hagenmuller, P. (1984) A simple model for predicting the electronic configuration of a d^n ion in a tetragonally distorted octahedral environment. *Proc. Ind. Acad. Sci.*, 93(3), 313-320.

[29] Chezhina, N. V., Akimenko, Yu. V. (1995) Sostoyanie atomov kobal'ta(III) v tverdykh rastvorakh YCaCo$_x$Al$_{1-x}$O$_4$.[In Russian] (State of cobalt(III) atoms in YCaCo$_x$Al$_{1-x}$O$_4$ solid solutions), *Zh. Obshchei Khimii*, 65(6), 902-905.

[30] *Advanced nanomaterials for Catalysis and energy*, (2019), Sadykov, V. Ed., Elsevier. 252-257.

[31] Rabanal, M. E., Varez, A., Levenfeld, B., Torralba, J. M. (2003) Magnetic properties of Mg-ferrite after milling process. *J. Mater. Proc. Tech.*, 143-144, 470-474.

[32] Spassky, D., Omelkov, S., Magi, H., Mikhailin, V., Vasil'ev, A., Krutyak, N., Tupitsina, I., Dubovik, A., Yakubovskaya A., Bel'sky, A. (2014) Energy transfer in solid solutions Zn$_x$Mg$_{1-x}$WO$_4$, *Optical Materials*, 36(10), 1660-1664.

[33] Burdett, J. K., Price, G. D., Price, S. S. (1982) The crystal field theory in determining the structure of spinels. *J. Am. Chem. Soc.*, 104, 92-95.

[34] Chezhina, N. V., Savel'eva, N. V. (1990) Magnitnaya otsenka raspredeleniya atomov Co(II) I Ni(II) v tverdykh rastvorakh M$_x$Mg$_{1-}$

$_x$Al$_2$O$_4$ so strukturoi shpineli [in Russian] (magnetic estimation of Co(II) and Ni(II) atom distribution over the M$_x$Mg$_{1-x}$Al$_2$O$_4$ solid solutions with spinel structure), *Zh. Neorg. Khim.*, 35(11), 2874-2876.

[35] Reznitskii, L. A. (1993) Energii izmeneniya koordinatsii kationov v oksidakh I sulfidakh [in Russian] (Energies of changing the cation coordination in oxides and sulfides), *Zh. Fiz. Khim.*, 67(12), 2379-2382.

[36] Chezhina, N. V., Galanova, I. V. (1994) Raspredelenie atomov khroma(III) I obmennye vzaimodeistviya v tverdykh rastvorakh MgCr$_x$Al$_{2-x}$O$_4$ [in Russian] (Distribution of chromium(III) atoms and exchange interactions in the MgCr$_x$Al$_{2-x}$O$_4$ solid solutions), *Zh. Neorg. Khimii*, 39(4), 582-584.

[37] Brach, B. Ya., Gorelova, A. V., Zvereva, I. A., Chezhina, N. V, (1985) Magnetokhimichskoe issledovanie tverdykh rastvorov ortogermanatov magniya soderzhashchikh elementy gruppy zheleza [in Russian] (Magnetochemical study of solid solutions of magnesium orthogermanates containing elements of iron group), *Zh. Neorg. Khim.*, 30(2), 301-305.

BIOGRAPHICAL SKETCHES

Natalia V. Chezhina

Affiliation: Sankt-Petersburg state university.

Education: Dr. of chemical sciences, professor.

Business Address: Universitetskaya nab. 7/9, Sankt-Petersburg, 199034, Russia.

Research and Professional Experience: Solid state chemistry, magnetochemistry of solids.

Professional Appointments: professor of Inorganic chemistry department.

Honors: Honored worker of high education.

Publications from the Last 3 Years:

Chezhina, N., D. Korolev, R. Bubnova, Y. Biryukov, O. Glumov, V. Semenov. Electronic structure of diluted SrFexTi1-xO3-δ solid solutions. *Journal of Solid State Chemistry* 274 (2019) 259–264.

Electronic Structure of Materials Challenges and Developments. N. Chezhina, D. Korolev Eds. Pan Stanford Publising. 2019.

Evarestov, R. A., A. V. Bandura, M. D. Sapova, D. A. Korolev, N. V. Chezhina. Parameterization of dilute Ising model for iron-containing lanthanum gallate and aluminate solid solutions based on first-principles calculations. *Solid State Ionics*. 2020. Vol. 348. P. 115283.

Ponomareva, E. A., A. V. Fedorova, and N. V. Chezhina. Magnetic Susceptibility of La1–yCeyAlO3 Solid Solutions. *Russian Journal of General Chemistry*, 2017, Vol. 87, No. 11, pp. 2730–2732

V. A. Sadykov, M. S. Koroleva, I. V. Piir, N. V. Chezhina, D. A. Korolev, P. I. Skriabin, A. V. Krasnov, E. M. Sadovskaya, N. F. Eremeev, S. V. Nekipelov, V. N. Sivkov. Structural and transport properties of doped bismuth titanates and niobates. *Solid State Ionics*. 2018. V. 315. P. 33-39.

Zhuk, N. A., Chezhina, N. V., Belyy, V. A., Makeev, B. A., Miroshnichenko, A. S., Beznosikov, D. S., Nekipelov, S. V., Sivkov, V. N., Yermolina, M. V. Influence of barium and strontium atoms on magnetic properties of iron-containing solid solutions Bi2MNb2O9 (M-Ba, Sr). *Journal of Magnetism & Magnetic Materials* Jan 2019, Vol. 469, p574-579.

Zhuk, N. A., Chezhina, N. V., Belyy, V. A., Makeev, B. A., Yermolina, M.V., Miroshnichenko. Magnetic susceptibility of solid solutions Bi2SrNb2−2xFe2xO9−δ. *Journal of Magnetism & Magnetic Materials*. Apr2018, Vol. 451, p96-101

Dmitry A. Korolev

Affiliation: Sankt-Petersburg state university.

Education: PhD, Associate Professor.

Business Address: Universitetskaya nab. 7/9, Sankt-Petersburg, 199034, Russia.

Research and Professional Experience: X-ray diffraction, Solid state chemistry, magnetochemistry of solids.

Professional Appointments: associate professor of Inorganic chemistry department.

Publications from the Last 3 Years:
Chezhina, N., D. Korolev, R. Bubnova, Y. Biryukov, O. Glumov, V. Semenov. Electronic structure of diluted $SrFexTi1-xO3-\delta$ solid solutions. *Journal of Solid State Chemistry* 274 (2019) 259–264.
Electronic Structure of Materials Challenges and Developments. N. Chezhina, D. Korolev Eds. Pan Stanford Publising. 2019.
Evarestov, R. A., A. V. Bandura, M. D. Sapova, D. A. Korolev, N. V. Chezhina. Parameterization of dilute Ising model for iron-containing lanthanum gallate and aluminate solid solutions based on first-principles calculations. *Solid State Ionics*. 2020. Vol. 348. P. 115283.
Korolev, D. A., Chezhina, N. V., Glumov, O. V. Synthesis, Interatomic Interactions, Structure, and Conductivity of Lanthanum Gallate Doped with Nickel and Magnesium. *Russian Journal of General Chemistry*, 89(6), (2019), с. 1129-1135.
Plekhanov, S., Kuzmin, A. V., Tropin, E. S., Korolev, D. A., Ananyev, M. V. New mixed ionic and electronic conductors based on $LaScO_3$: Protonic ceramic fuel cells electrodes. 2020. *Journal of Power Sources,* 449, p. 227476.

Sadykov, V. A., M. S. Koroleva, I. V. Piir, N. V. Chezhina, D. A. Korolev, P. I. Skriabin, A. V. Krasnov, E. M. Sadovskaya, N. F. Eremeev, S. V. Nekipelov, V. N. Sivkov. Structural and transport properties of doped bismuth titanates and niobates. *Solid State Ionics*. 2018. V. 315. P. 33-39.

In: An Introduction to Electronic Structure ... ISBN: 978-1-53618-411-2
Editor: Nadia T. Paulsen © 2020 Nova Science Publishers, Inc.

Chapter 3

MATHEMATICAL MODELING OF ELECTRONIC STRUCTURE OF SOME NANOMATERIALS

Azad A. Bayramov[1,*] *and Arzuman G. Gasanov*[2]

[1]Institute of Control Systems of the Azerbaijan National Academy Sciences, Baku, Azerbaijan Republic
[2]Armed Forces War College of the Azerbaijan Republic, Baku, Azerbaijan Republic

ABSTRACT

This review has been devoted to the mathematical modeling and investigation of the electronic structure of some nanomaterials, composite materials, and graphene by using the Parameterized Model number 3 (PM3) semi-empirical method. One of the variants of the molecular orbitals method - the semi-empirical Wolfsberg– Helmholz method was used to investigate the properties of the nanoparticles. For construction of molecular orbitals Ag_{16} are used 5s-, 5py-, 5pz-, and 5px- valence Slater atomic orbitals of silver atoms. The analytic expression of the basis Slater

---

[*] Corresponding Author's E-mail: azad.bayramov@yahoo.com.

functions was defined. The orbital energies, ionization potential, the total electronic energy, and effective charge of atoms of silver nanoparticles were calculated by the solution of equations of the molecular orbitals method. The possibility of using the simple computer program developed in Delphi Studio working undo MS Windows OS for carrying out the quantum mechanical calculation of the electronic structure of nanoparticles has been investigated. The theoretical methodology is described for the realization of this simple computer. The numerous quantum mechanical calculations show that this computer program works correctly and it is useful for use based on Slater Atomic Orbitals. The electronic structure of the gold nanoparticles was investigated by the semi-empirical Wolfsberg – Helmholz method. As the atomic orbitals used 6s-, $6p_y$-, $6p_z$- and $6p_x$- Slater atomic orbitals of gold atoms. The orbital energies, potential ionization, the total electronic energy, and effective charge of atoms of gold nanoparticles were calculated. The results indicate that the gold nanoparticles are soft, electrophile, and conductive material. Theoretical models of shockproof composite materials based on two-layer graphene and multilayer polyvinylidene fluoride $C_{124}H_{40} + n(H-(C_2H_2F_2)_5-H)$ (n = 1,2,…,8) are constructed. The electronic structure is studied using the semiempirical PM3 method that is one of the options of the molecular orbital method. The orbital energies, ionization potentials, total electron energies, strength, and other properties of the considered material are calculated based on the theoretical models. The outlooks for application of these materials in the military field for manufacturing super strong and lightweight flak jackets are considered.

Keywords: electronic structure, nanomaterials, Parameterized Model 3, Slater function, atomic orbitals

INTRODUCTION

This review based on our scientific articles has been devoted to the mathematical modeling and investigation of the electronic structure of some nanomaterials, composite materials, and graphene by using the molecular orbitals method. One of the variants of the molecular orbitals method - the semi-empirical Wolfsberg–Helmholz method was used to investigate the properties of the silver nanoparticles. For construction of molecular orbitals of Ag_{16} are used 5s-, 5py-, 5pz-, and 5px- valence Slater atomic orbitals of silver atoms. The analytic expression of the basis Slater functions (Slater 1960, 31-50) was defined. The orbital energies, ionization potential, the total electronic energy, and effective charge of atoms of silver nanoparticles were calculated by the solution of equations of the molecular orbitals method. The results indicate that the Ag_{16} nanoparticles are soft, electrophile, and stabile semi-conductive material. For carrying out the quantum mechanical calculation of the electronic structure of nanoparticles the computer program is developed in Delphi Studio working under MS Windows.

Also, the electronic structure of the gold nanoparticles was investigated by the semi-empirical Wolfsberg – Helmholz method. Molecular orbitals are represented as a linear combination of valence atomic orbitals of the atoms of the nanoparticle. As the atomic orbitals used 6s-, $6p_y$-, $6p_z$- and $6p_x$- Slater atomic orbitals of gold atoms. The exponential parameters of Slater functions were calculated and defined the analytic expression of the basic functions. The orbital energies, potential ionization, the total electronic energy, and effective charge of atoms of gold nanoparticles were calculated. The results indicate that the gold nanoparticles are soft, electrophile, and conductive material.

In this issue there have been presented the results of mathematical modeling of the molecular structure of polymer matrix hybrid micro- and nanocomposites materials having three phases: polyvinylidene fluoride (PVDF) molecule, the microparticle of $Pb_2(ZrTiO_6)$ piezoelectric and nanoparticle of $(SiO_2)_6$ dielectric ($PVDF + Pb_2(ZrTiO_6)+(SiO_2)_6$) by using of Parameterized Model number 3 (PM3) semi-empirical method. Theoretical models of shockproof composite materials based on two-layer graphene and multilayer polyvinylidene fluoride $C_{124}H_{40} + n(H-(C_2H_2F_2)_5-H)$ ($n = 1\div8$) are constructed. The electronic structure is studied using the semi-empirical PM3 method that is one of the options of the molecular orbital method. The outlooks for application of these materials in the military field for manufacturing super strong and lightweight flak jackets are considered.

THE COMPUTER PROGRAM FOR THE STUDY OF NANOPARTICLES IN BASIS OF SLATER ATOMIC ORBITALS

The computer program is developed in Delphi Studio working under MS Windows for carrying out of the quantum mechanical calculation of the electronic structure of some nanoparticles (Au_{16}, Ag_{16}, and $(CdS)_9$) (Gasanov et al., 2016). The calculations carried out by the Wolfsberg-Helmholz (WH) method (Maharramov et al., 2016; Ramazanov et al., 2014) based on Slater Atomic Orbitals (SAOs). It is known, that WH method is one of the simple semi-empirical variants of the molecular orbital (MO) method (Fedorov et al., 2006; Shembelov et al., 1980). In a method it is considered that each electron in a nanoparticle goes independently of other electrons in a certain effective field, which created by the nucleus and other electrons (Magerramov et al., 2010). The states of electrons in a nanoparticle are described by the one-electronic wave function called the molecular orbital. It is a multicenter function such includes distances from electron to various nuclei. There are various variants of searching for molecular orbital. One of them - search molecular orbitals U_i as a linear combination of atomic orbitals of the atoms, which are included in a nanoparticle (method MO LCAO) (Maharramov et al., 2016):

$$U_i = \sum_{q=1}^{m} c_{qi}\chi_q \tag{1}$$

Here c_{qi} are unknown coefficients, χ_q are atomic orbitals, chosen as basic functions. In this part of the work, the real SAOs are used as basic functions (Maharramov et al., 2016).

Usually in quantum-mechanical calculations of the electronic structure of nanoparticles limited to consideration of valence electrons of atoms and molecular orbitals are represented as linear combinations SAOs of these valence electrons. Coefficients c_{qi} in the formula (1) are found by the solution of the following system of equations;

$$\sum_q (H_{pq} - \varepsilon_i S_{pq})c_{qi} = 0, \tag{2}$$

Here the following designations are used:

$$H_{pq} = \int \chi_p^* \hat{H}ef \chi_q dV \qquad (3)$$

$$S_{pq} = \int \chi_p^* \chi_q dV \qquad (4)$$

S_{pq} is the overlap integrals between atomic orbials χ_p and χ_q. $\hat{H}ef$ is an effective Hamilton operator, moving in the certain effetive field:

$$\hat{H}ef = -\frac{1}{2}\nabla^2 + U(r) \qquad (5)$$

Expression of the effective Hamilton operator is unknown, therefore the values of matrix elements H_{pq} cannot be calculated analytically and it is estimated with various methods. The values of ionization potentials of atoms are used in the WH method for the estimation of matrix elements H_{pq}. Diagonal elements H_{qq} of this matrix take to equal potentials of ionization of corresponding valence states of atoms. Non-diagonal elements calculated using the expression given in (Maharramov et al., 2016; Ramazanov et al., 2014).

As it can be seen from formula (2) for quantum mechanical calculation of nanoparticles by method WH it is necessary to know the values of overlap integrals. In this work for calculation of overlap integrals, given in (Pashaev, 2009) the formulas are used. It is necessary to enter n, l, m quantum numbers of corresponding atomic orbitals, values of ξ exponential parameter of Slater type atomic orbitals (STOs) and Cartesian coordinates of atoms in a molecular system of coordinates for carrying out computer calculations of overlap integrals under these formulas. Knowing values of matrix elements H_{pq} and S_{pq}, it is possible to solve the system of the equations (2) and to calculate orbital energy ε_i, total electronic energy $E = \sum_i n_i \varepsilon_i$, values of ionization potential I_p of nanoparticle and values of c_{qi} coefficients in WH approach. Using the values of c_{qi} coefficients it is possible to calculate effective charges of atoms (q_A) and overlap populations (S_{Op}) in a nanoparticle by the following formula given in (Dmitriev, 1986):

$$q_A = n_A^0 - \sum_i n_i \sum_{q \in A} |c_{qi}|^2 \tag{6}$$

$$S_{op} = \sum_i n_i \sum_{p \in A} \sum_{q \in B} c_{pi} \cdot c_{qi} \cdot S_{pq} \tag{7}$$

Here: n_A^0 is a positive charge of the atom (for atoms of gold and silver $n_A^0 = 1$, for atoms of Cd $n_A^0 = 2$, for atoms of S $n_A^0 = 6$), n_i is the number of electrons on i-th molecular orbital. Summation over i is carried out on molecular orbital occupied by electrons. Using the described methodology developed computer program in Delphi Studio, working under MS Windows for carrying out of the quantum mechanical calculation of the electronic structure of the nanoparticles.

DECLARATION OF SOME PROGRAM

DKCTVK is a definition of the data;
nv is a number of basic functions SAOs;
mo is a number of molecular orbitals;
me is a number of atoms in a nanoparticle;
eos is a number of molecular orbitals, occupied by electrons;
nq, lq and mg are a principal, an orbital and a magnetic quantum numbers of basic SAOs, respectively;
tc is a type of the centers;
ch is a serial number of atoms included nanoparticles;
zci are the values of exponential parameter corresponding basic SAOs;
xc, yc, zc are the Cartesian coordinates of atoms in molecular system of coordinates;
DKBANTK is a set of functions and procedures for calculation of overlap integrals based on SAOs;
OVER calculates overlap integrals;
CLEBS calculates next coefficients:

$$(j_1j_2m_1m_2|j_1j_2JM) = \delta_{M,m1+m2} \times$$
$$\times \left[\frac{(2J+1)^2 F_{j1+j2}(j1 2 + j - 1.0)}{(2j_1+1)(2j_2+1)F_{j1-j2+J}(j1 2 + j + 1.0)} \times \right.$$
$$\left. \times \frac{F_{J+M}(2J,0)}{F_{j2-j1+J}(j1+j2+j+1.0)F_{j1+m1}(2j1,0)F_{j2+m2}(2j2,0)} \right]^{1/2} \times \quad (8)$$
$$\times \sum_n (-1)^n F_n(j_1 + j_2 - J, 0) F_{j2+m2-n}(J + M, 0) F_{j1-m1-n}(J - M, 0)$$

Here, $F_m(N, 0) = \frac{N!}{m!(N-1)!}$

$$n = \begin{Bmatrix} 0 \\ j_2 + m_2 - (J + M) \\ j_2 - m_2 - (J - M) \end{Bmatrix}_{max,...,} \begin{Bmatrix} j_1 + j_2 - J \\ j_2 + m_2 \\ j_1 - m_1 \end{Bmatrix}_{min}$$

FS calculates coefficients of binominal product

$$F_m(N, N') = \sum_{\sigma = \frac{1}{2}[(m-M)+|m-N|]}^{\min(m, N')} (-1)^\sigma F_{m-\sigma}(N, 0) F_\sigma(N', 0)$$

GB calculates coefficients

$$g_{\alpha\beta}^q(l\lambda, l'\lambda'; \Lambda) = g_{\alpha\beta}^0(l\lambda, l'\lambda'; \Lambda) F_q(\alpha + 2\Lambda - \lambda, \beta - \lambda')$$

Here: $g_{\alpha\beta}^0(l\lambda, l'\lambda'; \Lambda) = \sum_{i=1}^{\Lambda}(-1)^i F_j(\Lambda, 0) K_{\alpha+2\Lambda-2i}^{l\lambda} K_\beta^{l'\lambda'}$,

$$K_\beta^{l,\lambda} = (-1)^{\frac{1}{2}(l-\beta)} \sqrt{\frac{(2l+1)(l-\lambda)!}{2(l+\lambda)!}} \frac{(l+\beta)!}{2^l \left[\frac{1}{2}(l-\lambda)\right]! \left[\frac{1}{2}(l+\lambda)\right]! (\beta-\lambda)!}$$

DIS calculates distance between nucleuses of atoms;
DLM calculates coefficients

$$d_{m0}^0(t) = l! \, [(l+m)!\,(l-m)!]^{\frac{1}{2}} \sum_s \frac{(-1)^s \left[\frac{1}{2}(1-t)\right]^{s-\frac{m}{2}} \left[\frac{1}{2}(1+t)\right]^{l+\frac{m}{2}-s}}{s!\,(l-s)!\,(l-m)!\,(l+m-s)!};$$

BS calculates values of function

$$B_n(\beta) = \int_{-1}^{1} v^n e^{-\beta v} dv$$

and also keeps in one-dimensional massive;

ASN calculates values of function

$$A_n(p) = \int_{1}^{\infty} \mu^n e^{-p\mu} d\mu$$

and also keeps in one-dimensional massive;

AF calculates values of function

$$Q_{N\,N'}^q(p,t) = \int_{1}^{\infty} \int_{-1}^{1} (\mu+v)^N (\mu-v)^{N'} e^{-p\mu-pt} d\mu dv$$

RC calculates values of function

$$N_{nn'}(p,t) = \frac{(1+t)^{n+\frac{1}{2}} 1 - t^{n'+\frac{1}{2}}}{\sqrt{(2n)!\,(2n')!}} \cdot p^{n+n'+1}$$

OVERRUN calculates overlap integrals arising in research of properties of nanoparticles and preservation of results.

RESULTS OF CALCULATIONS

6s- 6px-, 6py- and 6pz- valence Au STOs of Au atoms, 5s- 5px-, 5py- and 5pz- valence orbitals of Ag atoms, 3s- 3px-, 3py- and 3pz- valence orbitals of S atoms, 5s- 5px-, 5py- and 5pz- valence orbitals of Cd atoms are used for the creation of molecular orbitals of nanoparticles. The analytical expressions of STOs given in (Maharramov et al., 2016; Ramazanov et al., 2014). By solving the equations (2) the values of orbital energies ε_i and the coefficient c_{qi} had been defined. The effective charges of atoms of nanoparticles and overlap populations can be calculated by using the values of coefficients cqi.

By using the values of orbital energies, the stability can be determined, and the electrical, mechanical, optical, and magnetic properties of nanoparticles can be calculated. The valence electrons of nanoparticles are placed in the lowest energetic levels two by two. The value of bandgap can be calculated as $E_g = \varepsilon_{LUMO} - \varepsilon_{HOMO}$. Here ε_{LUMO} is the energy of the lowest unoccupied molecular orbital and the ε_{HOMO} is the energy of the highest occupied by the valence electrons of molecular orbitals. If the value of E_g is in interval [0; 0,0251 eV, the material is conductive; if E_g is in interval (0.025; 6) eV, the material is semiconductive; if $E_g > 6$ eV the material is dielectric. The ionization potential is equal to ε_{HOMO} with the negative sign: $I_p = -\varepsilon_{HOMO}$. The strength of the material is calculated as $\eta = \frac{1}{2}E_g$. If $\eta > 1$ eV the material is strength, if $\eta < 1$ eV it is soft. When $\varepsilon_{LUMO} < 0$ the material is electrophile, when $\varepsilon_{LUMO} > 0$ it is nucleophile. The stability of nanoparticle can be calculated by formula $\Delta E = E_{np} - E_p$. Here E_{np} is the total electronic energy of nanoparticle, and E_p is the sum of total electronic energy of atoms in the nanoparticle. When $\Delta E > 0$ the material is not stable, but when $\Delta E < 0$ material is stable. The wavelength of photon under emission of the material can be calculated by the formula:

$$\lambda = \frac{c}{1.6 E_g} \times 10^{28} nm$$

Here: $c = 3 \cdot 10^8$ m/s is the speed of the light in vacuum, $h = 6.63 \cdot 10^{-34}$ C·s.

Table 1. Results of calculations for some nanoparticles based on SAOs

Nano-particles	E (amu.)	ΔE(amu)	I_P (eV)	E_g (eV)	Strength of nanoparticles	λ (mcm)
AU_{16}	-13.852	-0.329	16.648	0.405	0.202	3.1
AG_{16}	-15.027	-0.355	19.771	1.154	0.577	1.1
$(CdS)_9$	-39.103	-0.650	9.858	0.099	0.050	0.5

The results of calculations are given in Table 1. As seen from the table, the Au_{16}, Ag_{16}, $(CdS)_9$ nanoparticles are stable, soft, and semi-conductive materials.

THE STUDY OF GOLD NANOPARTICLES BASED ON THE SLATER FUNCTIONS

The Used Method

The gold nanoparticles have a wide range of applications due to their novel properties. These nanoparticles are using in the preparation of different transmitters, in electronics, in medicine the diagnostics of various diseases, in the chemical processes as catalysts and its application fields are expanding. For this reason, the study of the electronic structure of the gold nanoparticles (Figure 1) by quantum mechanics methods has great importance (Ashkarran, 2012; Liu et al. 2007). It is known that the sizes and the number of atoms in nanoparticles determine the structure and properties of nanoparticles. The size of nanoparticles which is consists of N atoms is given at the following formula (Liu et al. 2007).

$$D = \sqrt[3]{\frac{6MN}{\pi \rho N_A}} \quad (9)$$

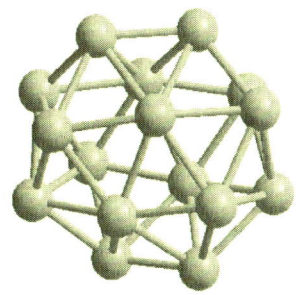

Figure 1. Electronic structure of the gold nanoparticle.

Here, N is the number of atoms, M- molar mass, ρ-material density, and N_A –Avogadro number. The calculated size of gold nanoparticles consisted of N = 16 atoms by the formula (1) is obtained D = 0.8 nm.

In the work (Pashaev et al. 2014) the electronic structure of the Au_{16} nanoparticles was investigated by the semi-empirical WH method. It is known that the WH method is a simple semi-empirical variant of the molecular orbital method (Shembelov et al. 1980; Fedorov et al. 2006; Minkin et al. 2010; Alieva et al. 2009). In MO the state of the electron is described with one-electron wave function so-called molecular orbital. Molecular orbitals are represented as a linear combination of valence atomic orbital of the atoms of the nanoparticles. Molecular orbitals are multicenter functions. Thus, the distances of an electron from a variety nucleus of atoms included in their expression. There are various ways to construct molecular orbitals. One of them is MO LCAO approximation. In this approximation, the molecular orbitals are written as linear combinations of valence atomic orbitals of atoms (1).

In this part of the work, the real STO's were used as basic functions. It is well known that the calculation of multicenter matrix elements over exponential type orbitals (ETO's) is the great importance for an accurate evaluation of problems in quantum chemistry and physics. Among the ETO's commonly used are the Gaussian type orbitals (GTO's) and STO's. The STO's represent the real situation for the electron density in the valence region, but are not so good nearer to the nucleus. Many calculations over the years have been carried out with STO's (Berlu et al.,

2005; Safouhi, 2004; Berlu et al. 2004; Guseinov, 2005; Guseinov, 2009). The real STO's functions are determined as

$$\chi_q \equiv \chi_{nlm}(\xi, \vec{r}) = \frac{(2\xi)^{n+\frac{1}{2}}}{\sqrt{(2n)!}} r^{n-1} e^{-\xi r} S_{lm}(\theta, \varphi) \tag{10}$$

$S_{lm}(\theta,\varphi)$ are the real spherical harmonics:

$$S_{lm}(\theta, \varphi) = \frac{1}{\sqrt{\pi(1+\delta_{m0})}} P_{l|m|}(\cos\theta) \begin{cases} \cos|m|\varphi, m \geq 0 \\ \sin|m|\varphi, m < 0 \end{cases} \tag{11}$$

Here: $P_{l|m|}(\cos\theta)$ are the normalized associated Legendre functions (Gradshteyn & Ryzhik, 1980) and n, l, m are the principal, orbital, and magnetic quantum numbers, ξ is an exponential parameter determined by formulas (Pashaev, 2014):

$$\gamma_i = \sum_{j \neq i}^{N} \left\{ 1 + \left[\frac{3n_j^2 - l_j(l_j+1)}{3n_i^2 - l_i(l_i+1)} \right]^2 \right\}^{-\frac{3}{2}} \tag{12}$$

$$\xi = \frac{Z-\gamma}{n} \tag{13}$$

Here: Z is an atom number. Usually in quantum mechanics calculations of the electronic structure of molecules, it is satisfied considering only the atomic orbitals of valence electrons. For the creation of molecular orbitals of gold nanoparticles, 4 valence atomic orbitals (6s, 6p$_y$, 6p$_z$, 6p$_x$) from each gold atoms are taken. Thus, 64 Slater's atomic orbitals were used. The analytic expressions of atomic orbitals are considered as follow:

$$\chi_1 = 6s(Au) = \frac{1.027405}{\sqrt{\pi}} \cdot r^5 e^{-2.599004r} \tag{14}$$

$$\chi_2 = 6p_y(Au) = \frac{1.316146}{\sqrt{\pi}} \cdot r^5 e^{-2.481152r} \sin\theta \sin\varphi \tag{15}$$

$$\chi_3 = 6p_z(Au) = \frac{1.316146}{\sqrt{\pi}} \cdot r^5 e^{-2.481152r} \cos\theta \tag{16}$$

$$\chi_4 = 6p_x(Au) = \frac{1.316146}{\sqrt{\pi}} \cdot r^5 e^{-2.481152r} \sin\theta\cos\varphi \tag{17}$$

In the expressions of (14) – (17) r, θ and φ are spherical coordinates of an electron. Based on the formula (1) 64 molecular orbitals had established. The nanoparticle, which was created from 16 gold atoms, has valence electrons. They fill eight low energetic levels. The basic functions of other gold atoms are determined similarly. The unknown coefficients are found by solving the following system of equations (2), (3), and (4).

The quantities of H_{pq} are matrix elements of the effective Hamiltonian (5), for one electron moving in a molecule in some effective field independent from others. Thus, for the solution of equations system (2), i.e., for determinations of the orbitals energies ε_i and corresponding sets of C_{qi} coefficients, the numerical H_{pq}, and S_{pq} values must be known. However, H_{pq} values cannot be calculated exactly because the explicit expressions for the operator are unknown. Therefore, it is necessary to estimate them in various ways, one of which based on quantum chemical semi-empirical method VH. According to method VH, each diagonal matrix element H_{pq} are equal to an ionization potential according to valence state of the given atoms. The non-diagonal elements are defined by a ratio (Shembelov et al. 1980; Feodrov et al. 2006).

$$H_{pq} = 0.5 \cdot K \cdot S_{pq}(H_{pp} + H_{qq}) \tag{18}$$

expression calculates overlap integrals. Here: the meaning of coefficient K is established by theoretical from the condition of a minimum of energy or comparison with experimental data. As seen from (2) and (18), the expression for the implementation of quantum mechanical calculating by VH method it is important to know the value of overlap integrals in molecular coordination system. In this work, for the of STO's were used the expressions from given in (Guseinov et al. 1989; Guseinov, 2003; Guseinov et al. 1998; Pashaev, 2009).

Based on these expressions, for the calculating overlap integrals the *n*, *l* and *m* quantum numbers, exponential parameters of atomic orbitals ξ, and the Cartesian coordinates of atoms should be taken. To calculate of H_{pq} matrix elements the following value of potential ionization of 6s valence state of gold atoms have been used:

(6s|Au|6s) = -0.3389363 amu.

By knowing the value of H_{pq} and S_{pq} matrix elements, and solving the system equations (18), the value of orbital energies ε_i, the total electronic energy $E = \sum_i \varepsilon_i$, the ionization potential I_p, and coefficients C_{qi} can be found in the VH approach.

The numerical values of coefficients C_{qi} allow determining the effective charge q_a (in amu.) of an atom A in the molecule according to the MO LCAO method by the formula (Dmitriev, 1986).

$$q_A = n_A^o - \sum_i n_i \sum_{q \in A} |C_{qi}|^2 \qquad (19)$$

Here: n_A^o is a positive charge of the nuclear core of atom A (for the gold atoms $n_A^o=1$), n_i is the number of electrons in the i-th molecular orbitals. Summation for *i* is performed over the occupied molecular orbitals. The software for computations the numerical values of C_{qi}, orbital energies ε_i, total energy E, potential ionization I_p, and an effective charge of atoms has been designed in VH approach.

THE COMPUTER CALCULATIONS FOR AU$_{16}$ GOLD NANOPARTICLES BY THE WH METHOD

Below, the computer calculations of some properties of Au$_{16}$ nanoparticles by the Wolfsberg-Helmholz method have been presented:

a total electronic energy E = -6.339366 amu;

a potential ionization $I_p = 3.703389914$ eV.

Calculated orbital energies (amu.) are presented in Table 2:

Table 2. Calculated orbital energies for Au_{16} nanoparticles

i	ε_i	i	ε_i	i	ε_i	i	ε_i
1	-0.684817	17	0.000000	33	0.000000	49	0.016042
2	-0.661433	18	0.000000	34	0.000000	50	1.204702
3	-0.623058	19	0.000000	35	0.000000	51	1.229904
4	-0.598719	20	0.000000	36	0.000000	52	1.285250
5	-0.169649	21	0.000000	37	0.000000	53	1.303929
6	-0.158814	22	0.000000	38	0.000000	54	1.349486
7	-0.137098	23	0.000000	39	0.000000	55	1.407488
8	-0.136096	24	0.000000	40	0.000000	56	1.970694
9	-0.118915	25	0.000000	41	0.000000	57	2.249531
10	-0.027240	26	0.000000	42	0.000000	58	2.896924
11	-0.024011	27	0.000000	43	0.000000	59	3.378143
12	-0.019275	28	0.000000	44	0.000000	60	3.463406
13	-0.016488	29	0.000000	45	0.000000	61	3.665532
14	-0.012503	30	0.000000	46	0.000000	62	3.846329
15	-0.010608	31	0.000000	47	0.000000	63	3.914490
16	-0.000524	32	0.000000	48	0.000000	64	3.988562

Table 3. Effective charges of atoms and coordinates

Charge	Coordinates (amu.)		
	x	y	z
-0.477725	0.7273814	0.14770909	2.65187515
-0.556951	-1.87186795	0.56232986	2.58630971
-0.609782	0.3630383	-2.26437264	1.37450644
-0.477706	-2.20357127	-1.40494744	0.86809597
-0.489172	2.67199854	-0.86560465	1.07587674
-0.537255	2.24897258	1.75132253	1.07268487
-0.342177	-0.36228144	2.40525927	1.39682705
0.431552	0.00377388	0.05200688	-0.05143089
-0.590981	0.87839365	2.64291629	-0.99603353
-0.342212	-2.49680715	1.27442059	0.09783097
-0.498185	2.38512292	0.43716801	-1.35228525
-0.489191	-0.80251563	-2.70621679	-1.03866935
-0.537262	-2.39001153	-0.70620755	-1.75069312
-0.349636	1.78344705	-2.18175034	-1.03139997
-0.590996	-1.14737978	1.56980459	-2.22882881
-0.498192	0.21230642	-0.71383772	-2.67466599

Calculated the values of effective charges of Au_{16} atoms and coordinates are presented in Table 3 ($Z_A = 79$).

Starting from the lowest energy level, 16 valence electrons of Au_{16} nanoparticles are placed in levels two by two. The energy of the highest level which occupied by electrons equals the value of potential ionization with a negative sign, $I_p = -\varepsilon_8 = 3.703389914$ eV. The value of bandgap can be calculated as the difference in the energy lowest unoccupied molecular orbital $\varepsilon_{LUMO} = \varepsilon_9$ and the energy of the highest occupied molecular orbitals $\varepsilon_{HUMO} = \varepsilon_8$: $\varepsilon_{LUMO} - \varepsilon_{HUMO} = 0.4675225$ eV. It indicates that Au_{16} nanoparticles are conductors. The strength can be calculated as $\eta = \frac{1}{2}(\varepsilon_{LUMO} - \varepsilon_{HUMO}) = 0.23376125$ amu. Thus, $\eta < 1$ eV and Au_{16} nanoparticles are soft material. The energy of the lowest unoccupied molecular orbital is negative, sign, thus Au_{16} nanoparticles are electrophilic. The stability of Au_{16} nanoparticles can be expressed by the formula $\Delta E(Au_{16}) = E_{Au8} - 8 \cdot E_{Au2}$. Here, $\Delta E(Au_{16})$ is the parameter which identified the stability of Au_{16} nanoparticles. If the $\Delta E(Au_{16}) > 0$, then material is not stable, but if $\Delta E(Au_{16}) < 0$, then material is stable. E_{Au16} is the total energy of Au_{16} nanoparticles, E_{Au2} - is the total energy of Au_2 molecules. Due to $E_{Au16} = -6.339366$ amu, $E_{Au2} = -0.759462$ amu and $\Delta E(Au_{16}) = -0.26367$ amu, that is $\Delta E(Au_{16}) < 0$ and Au_{16} nanoparticles are stable.

INVESTIGATION OF SILVER NANOPARTICLES BASED ON SLATER FUNCTION

The silver nanoparticles have a wide range of applications such as in the preparation of different transmitters, in electronics, for diagnostics of various diseases in medicine, in the chemical processes as a catalyst and its application fields are expanding (Andrea et al., 2008). The study of the electronic structure of the nanoparticles by quantum mechanics methods has great importance (Ramazanov et al., 2014; Pashaev et al., 2014). The number of atoms in the nanoparticles determines the size-dependent structural and energetic properties of nanoparticles. The shape of Ag_{16} nanoparticles is considered as a sphere and the size of nanoparticles can be calculated by the following formula (9).

The calculated size of Ag_{16}, by the formula (9) is obtained $D \approx 0,8$ nm. The semi-empirical Wolfsberg – Helmholz method was used to investigate the properties of the silver nanoparticles (Maharramov et al., 2016). In MO method, the state of the electron is described with one-electron wave function so-called molecular orbital. Molecular orbitals can be represented as linear combinations of atomic orbitals of atoms of nanoparticles (1).

For an investigation of Ag_{16} nanoparticles, the real Slater type atomic orbitals (STO's) (10) were used as basic functions. In the quantum-mechanical investigation of the properties of molecules and nanoparticles, the exponential type orbitals (ETO's) has great importance (Guseinov, 2012; Santos et al., 2014). Gaussian type orbitals (GTO's) and STO's are the most commonly used ETO's. It is reasonable to use STO's in valence electronic approximation.

For the creation of molecular orbitals of Ag_{16} nanoparticles are taken 4 valence atomic orbitals 5s, $5p_y$, $5p_z$, $5p_x$ from each silver atoms. The analytic expressions of these atomic orbitals are considered as follows:

$$\chi_{5s}(1.992739, r) = \frac{0.5269031}{\sqrt{\pi}} \cdot r^4 e^{-1.992739}$$

$$\chi_{5px}(2.065968, r) = \frac{1.112997}{\sqrt{\pi}} \cdot r^4 e^{-2.065968} \sin\theta\cos\varphi$$

$$\chi_{5py}(2.065968, r) = \frac{1.112997}{\sqrt{\pi}} \cdot r^4 e^{-2.065968} \sin\theta\cos\varphi$$

$$\chi_{5pz}(2.065968, r) = \frac{1.112997}{\sqrt{\pi}} \cdot r^4 e^{-2.065968} \cos\theta$$

Ag_{16} has 16·1=16 valence electrons. They are situated in eight low energetic levels. The quantities C_{qi} are found by solving the system of (2)-(4) equations.

To calculate H_{pq} matrix elements we use the following values of the ionization potential of 5s and 5p valences state of silver atoms:

$(5s|Ag|5s) = -0.789736$ amu; $(5p|Ag|5p) = -0.278332$ amu.

By knowing the value of H_{pq} and S_{pq} matrix elements and solving the system equations (18), the value of orbital energies ε_i, the total electronic energy $E = \sum_i \varepsilon_i$, the ionization potential I_p, and coefficients C_{qi} can be found in the VH approach. The effective charge (in amu) of an atom A in the nanoparticle can be calculated by the formula (19). The calculated values of orbital energies are given in Table 4. The effective charges and coordinates of atoms are given in Table 5: a total electronic energy $E = -15.027638$ amu; an ionization potential $I_p = 19.7719938$ eV.

Table 4. The values of orbital energies of Ag$_{16}$

The values of orbital energies ε_i (a.u.)			
$i=1,...,16$	$i=17,...,32$	$i=33,...,48$	$i=49,...,64$
-1.170832	-0.361529	-0.255869	0.337688
-1.083916	-0.341670	-0.244131	0.381342
-1.066290	-0.338956	-0.214240	0.462554
-1.014616	-0.338876	-0.204089	0.536640
-0.837913	-0.327527	-0.137771	0.614872
-0.831932	-0.317396	-0.127481	0.627916
-0.781719	-0.316707	-0.073974	0.640551
-0.726601	-0.312671	-0.070470	0.752829
-0.684174	-0.312071	-0.064365	0.868786
-0.557607	-0.306468	0.002556	0.971381
-0.541771	-0.303170	0.005128	1.008168
-0.493868	-0.294944	0.058071	1.097059
-0.490898	-0.281399	0.060850	1.144841
-0.483505	-0.281077	0.067135	1.261422
-0.461802	-0.272556	0.185937	1.311763
-0.395196	-0.268143	0.190002	1.379747

The 16 valence electrons of Ag$_{16}$ nanoparticles are placed in the first eight energetic levels two by two. The ionization potential is equal to ε_8 with negative sign. $I_p = -\varepsilon_8 = 19.7719938$ eV. The value of band gap can be calculated as $E_g = \varepsilon_{LUMO} - \varepsilon_{HUMO}$. Here, $\varepsilon_{LUMO} = \varepsilon_9 = -18.61748619$ eV, is the energy of the lowest unoccupied molecular orbital and the $\varepsilon_{HUMO} = \varepsilon_8$ is the energy of the highest occupied molecular orbitals.

Table 5. Effective charges and coordinates of atoms of Ag$_{16}$

Effective charge of atom	Coordinates (amu)		
	X	Y	Z
0.257275	-5.03929687	-2.098277136	-2.31674846
0.257295	2.705200717	5.273547414	-0.180223252
0.257286	5.637206455	1.525236358	1.028937388
0.387377	2.580667716	1.011740873	-3.033484071
0.262163	4.730458807	-3.028967623	-0.790397164
0.280563	4.183666335	-2.054019733	4.15955342
0.387397	1.118094702	2.002600264	3.409842076
0.262116	-2.124922285	5.152529305	1.055620331
0.280535	-0.680717416	4.908508873	-3.803207622
0.387392	-3.602329655	0.378247732	1.940352661
0.262127	-0.723179579	-2.423045597	5.077885093
0.280534	1.53649913	-5.728612544	1.960251485
0.387382	-0.096300482	-3.392645561	-2.31674846
0.262172	-1.882432532	0.299653991	-5.343051569
0.280542	-5.03929687	2.873953328	-2.31674846
0.257269	-3.303337069	-4.700449944	1.468166606

So, $E_g = 1.154508$ eV. The strength of the material is calculated as $\eta = \frac{1}{2}(\varepsilon_{LUMO} - \varepsilon_{HUMO}) = 0.577254$ eV. As can be seen, Ag$_{16}$ nanoparticles are soft, semi-conductive material. ε_{LUMO} has a negative sign, therefore, Ag$_{16}$ are electrophilic.

The stability of Ag$_{16}$ nanoparticles can be expressed by the formula $\Delta E(Ag_{16}) = E_{Ag8} - 8 \cdot E_{Ag2}$. When the material is not stable then $\Delta E(Ag_{16}) > 0$, but when material is stable then $\Delta E(Ag_{16}) < 0$. $E_{Ag16} = -15.027638$ amu is a total energy of Ag$_{16}$ molecules, $E_{Ag2} = -1.833982$ amu is a total energy of Ag$_2$ molecules. $\Delta E(Ag_{16}) < 0$ that is, Ag$_{16}$ nanoparticles are stable.

The accuracy of the results that came from the paper has been checked by the test calculations of the Ag$_2$ molecule using the other methods. A comparison of the results of different methods for Ag$_2$ is given in Table 6.

As seen in Table 6, there are differences in the calculations. These differences occur because of the type, number of the basic functions, and the variety of the electrons, which were used in the calculations.

Table 6. Results of computer calculations for Ag_2 by different methods

Object and methods	Number of electrons using in calculations	ε_{HOMO}	ε_{LUMO}	I_p(eV)	E_g(eV)
Ag_2 - WH (STO's)	2	-24.952815	-16.769817	24.9528	8.182998
Ag_2 - Ab Initio (GTO's)	94	-4.425674	4.172199	4.4257	8.597873
Ag_2 Extended Hukkel (GTO's)	22	-9.114572	-6.630959	9.1146	2.483613

ELECTRONIC STRUCTURE OF GRAPHENE-POLYVINYLIDENE FLUORIDE COMPOSITE MATERIAL

Graphene is represented in the form of a monatomic layer of carbon atoms connected in a two-dimensional lattice by σ and π bonds in a state of sp2 hybridization. A graphene crystal structure is a hexagonal lattice, and the lattice constant is 0.246 nm (Katsnelson, 2012; Castro et al., 2009; Geim et al., 2007). Atoms of carbon in the graphene layer are bound by strong covalent bonds. Three of four valence electrons of an atom of carbon participate in the σ bond, and one valence electron is involved in the π bond. In the π state, the electronic cloud overlaps each other less than in the σ state. Therefore, π bonds break easily so that π electrons travel around all atomic nuclei inside the lattice. This provides the conductivity of graphene. Graphene possesses a high mechanical hardness of ~1 TPa (Novoselov et al., 2007). In addition, acoustic waves propagate in graphene three times faster than in steel, and it means that graphene would faster absorb and disperse the energy of acoustic waves produced, for example, by a bullet shot from a gun. Graphene efficiently slows down the bullet preventing its further intrusion. Therefore, there is a real prospect of making a superstrong and lightweight bulletproof flak jacket based on graphene sheets (Gasanov et al., 2019). Graphene sheets disperse kinetic energy: they are stretched in the form of a cone at the point of contact with the bullet and then crack and fall to pieces.

The fragments are also a threat to a person. This problem can be solved by application of a multilayer graphene coating (which sharply raises its cost) or its inclusion into the polymer composite structure. In our study, we have considered a composite structure based on the two-layer graphene and the multilayer polymer matrix. The two-layer graphene is distinguished by its high mechanical strength (Novoselov et al., 2006). This property provides numerous applications of the graphene-containing materials, especially in the military field. Therefore, the study of two-layer graphene-containing shockproof materials is of great interest. For the purpose of protection, polyvinylidene fluoride (PVDF) inclusion as a splinter proof polymer matrix in a composite polymer material was considered. From this viewpoint, the theoretical studies of a composite material based on the two-layer graphene and polymer matrix $C_{124}H_{40} + nPVDF$ is very interesting. In this paper, the results of calculations of theoretical models of composite material $C_{124}H_{40}+n(H–(C_2H_2F_2)_5–H)$ (n=1,2,...,8) based on the two-layer graphene and the multilayer polymer matrix, and examinations of the electronic structure of this material using the PM3 semi-empirical method (Ranjbartoreh, 2011) are presented (Gasanov, 2019). Orbital energy of electrons ε_i were calculated. Then, the total energies of electrons and the values of ionization potentials were calculated using the values of ε_i, and certain mechanical, electrical, and optical characteristics of this $C_{124}H_{40} + n(H–(C_2H_2F_2)_5–H)$ shockproof material were studied. Mathcad office, free evaluate version HyperChem Professional v7.5 and MS Excel were used for the calculations.

METHODS

The mathematical simulation of various characteristics of shockproof materials made on the basis of graphene, and their quantum-mechanical examination is of great importance (Zavodinsky et al., 2004;1999;2005).

These theoretical calculations are usually carried out by the molecular orbital method (MO) (Zavodinsky et al., 2015; Ramazanov et al., 2017; Maharramov et al., 2016; Ramazanov et al., 2014; Pashaev et al., 2014). It is known that the semi-empirical method PM3 is a simple version of the molecular orbital (MO) method (Stewart et al., 1989). The MO method supposes that each electron of a molecule moves independently in a certain effective field created by other electrons and molecule nuclei. The state of an electron in a molecule is described by a one-electron wave function termed molecular orbital (Gasanov, et al., 2016). These functions are multicentered. Thus, their description includes the electron distances from atomic nuclei. There are various options to find molecular orbitals. One of them, the method of finding molecular orbital (MO LCAO method) Ui, as a linear combination of atomic orbitals of the atoms composing the molecule (1). In this study, the Gaussian functions were used as atomic orbitals.

The coefficients by solving the following system of equations (2)-(4) were found. Since potential function $U(r)$ in (5) is explicitly unknown, we cannot calculate H_{pq} precisely. They can be estimated using certain experimental parameters. The diagonal elements of matrix H_{pq} are taken equal to ionization potentials of the corresponding valence states of atoms with the opposite sign

$(1s|H|1s) = -0.499786 \ aum,$
$(2s|C|2s) = -0.772096 \ aum,$
$(2p|C|2p) = -0.419161 \ aum$

Nondiagonal elements of H_{pq} can be calculated in various approximations (Zavodinsky et al., 1999). The system of equations (2) is a system of linear homogeneous equations. Solving this system, we find numerical values of ε_i and c_{qi}. Using the values of ε_i we can calculate the values of the total energy of an electron and ionization potential, and examine the material properties.

In earlier papers, the PM3 semiempirical method was applied to study electronic structures of various materials (Gasanov et al., 2019). The calculations show that in C–C linear chains and in graphene-like clusters, irrespective of the size, equilibrium distances C–C are 0.13 nm (Zavodinsky et al., 1999). In this study, the theoretical models of composite material based on the two-layer graphene $C_{124}H_{40}$ and the multilayer polymer (PVDF) matrix $C_{124}H_{40} + (H- C_{10} + H_{10} + F_{10}–H)$ are calculated and plotted (Figure 2).

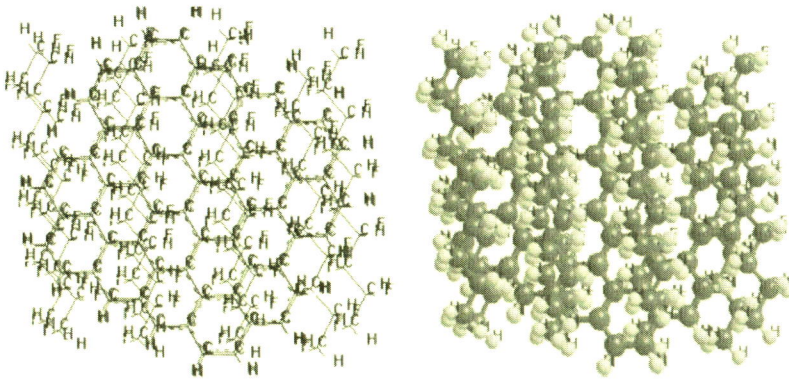

Figure 2. The counted models of electronic structure of a fragment of two-layer graphene in polymer matrix $C_{124}H_{40}$ + 2PVDF.

The following parameters were used: molecular mass of two-layer graphene $C_{124}H_{40}$ = 1529.6444 amu (Novoselov et al., 2006), molecular mass of PVDF = 322.186 amu. Distances C–C were assumed 0.13 nm, and further calculations were carried on using the PM3 method. The calculations were carried out under the following conditions: invariance of the molecular mass of two-layer graphene $C_{124}H_{40}$ and molecular mass PVDF incremented by at each step.

For each object, electrons are adjusted by two at energy levels, starting from the lowest level. Energy levels occupied by electrons of the top-most molecular orbitals ε_{HOMO} and the lowest empty molecular orbitals ε_{LUMO} were determined. The following parameters were calculated: ionization potential $I_p = -\varepsilon_{HOMO}$, the band gap width $E_g = \varepsilon_{LUMO} - \varepsilon_{HOMO}$ and the strength (hard or soft material) $\eta = 0.5E_g$. At $\eta < 1$ eV the material is considered soft, and at $\eta > 1$ eV, the material is considered hard.

The wavelength of a radiated quantum can be calculated by formula:

$$\lambda = \frac{c \cdot h \cdot 10^{28}}{1.6 \cdot E_g} \ (nm)$$

Here: h is the Planck constant, c is the velocity of light in vacuum, E_g is the band gap width in eV.

The material stability is calculated by the formula $\Delta E = E - \sum_A E_A$. Here E is the total energy of the system, E_A is the total energy of a free atom A included in the system, and ΔE is the parameter describing the system stability. At $\Delta E > 0$, the material is considered unstable, and at $\Delta E < 0$, the material is considered stable.

Table 7. Results of calculation of two-layer graphene characteristics in multilayer polymer matrix $C_{124}H_{40}$ + nPVDF (n =1, 2, 3, 4, 5, 6, 7, 8) using PM3 method

object	ε_{HOMO} amu	ε_{LUMO} amu	E, amu	ΔE, amu	I_p (eV)	E_g (eV)	η, eV	λ, nm
PVDF	−12.508	0.621	−212.19	−5.73	12.5	13.1	6.565	94.68
$C_{124}H_{40}$	−9.449	1.163	−551.95	−37.85	9.5	10.6	5.306	117.14
$C_{124}H_{40}$ +1PVDF	−9.038	1.198	−762.35	−41.79	9.0	10.2	5.118	121.44
$C_{124}H_{40}$ +2PVDF	−9.468	0.815	−973.9	−46.89	9.5	10.3	5.142	120.89
$C_{124}H_{40}$ +3PVDF	−8.882	0.962	−1184.9	−51.44	8.9	9.8	4.922	126.28
$C_{124}H_{40}$ +4PVDF	−9.183	0.742	−1396.9	−56.20	9.2	9.9	4.962	125.26
$C_{124}H_{40}$ +5PVDF	−9.345	0.545	−1607.6	−61.23	9.3	9.9	4.945	125.68
$C_{124}H_{40}$ +6PVDF	−9.238	0.637	−1819.5	−66.63	9.2	9.8	4.938	125.86
$C_{124}H_{40}$ +7PVDF	−7.906	0.502	−2028.9	−69.56	7.9	8.4	4.204	147.86
$C_{124}H_{40}$ +8PVDF	−7.908	0.604	−2242.1	−76.32	7.9	8.5	4.256	146.05

Based on the considered models of composite material $C_{124}H_{40}$ + 2(H–$(C_2H_2F_2)_5$–H), in the case, when the distance between graphene layers is 0.142 nm, the following parameters were calculated: orbital energies of electronic structures, ionization potentials, total energies, hardness, stability parameters, band gap width, and the wavelengths of the emitted photons. The results of calculation are presented in Table 7. Apparently, the material is electrophilic, dielectric, and stable.

Let us calculate the value of hardness per unit of elongation (Young modulus) for the considered composite material $C_{124}H_{40}$ + 2(H–$(C_2H_2F_2)_5$–

H). It is known that the force of graphene rupture can be calculated by formula:

$$F \approx \frac{E_b}{r}$$

Here E_b is the binding energy of the considered two layers of graphene + polyvinylidene fluoride, r is the distance of the bond between the atoms

$$E_b = 124E_c + 40E_H - E_{2graphene+PVDF}.$$

E_c, E_H and $E_{2graphene+pvdf}$ are the total energies of carbon, hydrogen and two-layer graphene + polyvinylidene fluoride calculated by the PM3 method (Gasanov et al., 2019); E_b is the binding energy of the two-layer graphene + polyvinylidene fluoride calculated by the PM3 method

$E_c = 1.74 \cdot 10^{-17}$ J; $E_H = 0.21 \cdot 10^{-17}$ J; $E_{2graphene+pvdf} = 2.41 \cdot 10^{-15}$ J
$E_b = 1.849 \cdot 10^{-16}$ J; $r = 0.13 \cdot 10^{-9}$ m

Table 8 shows that the highest value of the hardness parameter $\eta = 5.118$ eV is observed in material $C_{124}H_{40} + 2(H-(C_2H_2F_2)_5-H)$.

Table 8. Results of calculation of values of the shear modulus, the bulk elasticity modulus, the Young modulus, and hardness modulus of graphene, two-layer graphene, and two-layer graphene with a multilayer polymer matrix (PVDF) using PM3 method

Materials	Shear modulus	Bulk elasticity modulus (GPa)	Young modulus (GPa)	Hardness modulus (GPa)
Graphene	312	870.9	836.1	130
Two-layer graphene	335	935.3	897.9	139.6
Two-layer graphene with PVDF	549.5	1534	1473	228.9

Let substitute these calculated values into formula (1) and obtain the value of hardness per unit of elongation (Young modulus) for composite material

$$C_{124}H_{40} + 2(H-(C_2H_2F_2)_5-H)$$

$$Y = \frac{F}{S} = \frac{E_b}{r \cdot S}$$

Here S is the area of the graphene fragment cross section consisting of 62 carbon atoms 12C. Finally, the value of the Young modulus of the considered composite material is obtained.

If the binding energy of single-layer graphene is $E_b = 0.79 \cdot 10^{-16}$ J, then $Y = 629.1$ GPa, and if the binding energy of two-layer graphene is $E_b = 1.128 \cdot 10^{-16}$ J, then $Y = 897.9$ GPa, i.e., the Young modulus of two-layer graphene is 1.4 times larger, than that of the single layer material. The Young modulus $Y = 1.473$ TPa of two-layer graphene with the composite material is 2.3 times larger, than that of the single-layer material.

In order to calculate the shear modulus, bulk modulus, and hardness modulus of graphene, two-layer graphene, and two-layer graphene with composites, the following formulas (Openov et al., 2017; Gasanov, et al., 2019) has been used

$$G = \frac{Y}{2(1+v)}, K = \frac{Y}{3(1-2v)}, HN = G \cdot A e^{-BT}$$

Here $v = 0.34$ is the Poisson ratio (Jiang et al., 2016), Y is the Young modulus, G is the shear modulus, K is the bulk modulus of elasticity, HN is the hardness modulus, $A = 0.807$, $B = 2.204 \cdot 10^{-3}$, and $T = 300$ K. The calculated results are presented in Table 2. Hence, the considered composite material based on two-layer graphene $C_{124}H_{40} + 2(H-(C_2H_2F_2)_5-H)$ is very hard and lightweight: $\rho \approx 1.7$ g/cm^3. This is why it can be used to manufacture various hard and, at the same time, lightweight materials, in particular, flak jackets.

CONCLUSION

Thus, the semi-empirical Wolfsberg–Helmholz method was used to investigate the properties of the nanoparticles. STO's are used as atomic orbitals. The computer calculations were carried out by computer program Delphi studio system under the Windows operating system. The results of the calculations indicate that STO's are useful in the investigation of the properties of multielectron systems (molecules, nanoparticles).

The electronic structures of the gold nanoparticles were investigated by the semi-empirical WH method based on Slater functions. The orbital energies, the ionization potential, the total energy of electrons, and the effective charge of atoms of gold nanoparticles were calculated. The results of calculations show that gold nanoparticles are soft, electrophile, and conductive material, and the use of Slater functions in the study and application of nanosystems is appropriate.

The semi-empirical WH method was used to investigate the properties of the silver nanoparticles. The results of the calculations indicate that STO's are useful in the investigation of properties of nanoparticles in valence electronic approximation. The orbital energies, ionization potential, the total electronic energy, and effective charge of atoms of silver nanoparticles were calculated. The results of calculations show that silver nanoparticle is soft, electrophile, and stabile semi-conductive material.

The electronic structure of shockproof material $C_{124}H_{40} + n(H-(C_2H_2F_2)_5-H)$ ($n = 1,2,...,8$) based on two-layer graphene in the multilayer polymer matrix is studied applying the PM3 method and theoretical models are plotted. Based on the models, assuming the C–C distance to be 0.13 nm, and the distance between the graphene layers to be 0.142 nm, the values of orbital energy of electronic structures, ionization potentials, total energies, hardness, stability parameters, bandgap width, wavelengths of the emitted photons, and strength were calculated for these materials. The calculations have shown that the considered material based on two-layer graphene is electrophilic, dielectric, and stable and is 1.761 times stronger than single-layer graphene. The value of hardness (strength) of the strongest model of the composite material based on two-layer graphene $C_{124}H_{40}$ (H − C_{10} + H_{10} + F_{10} − H) (model $N = 2$) was found to be 228.9 GPa.

REFERENCES

Alieva, R.A., Pashaev, F.G., Gasanov, A.G., Mahmudov, K.T. 2009. Quantum chemical calculations of the tautomeric forms of azo derivatives of acetylacetone and determination of the stability constants of their complexes with rare-earth metals. *Russian Journal of Coordination Chemistry.* 35 (4), 241-246.

Andrea, R. Tao, Susan, Habas, and Peidong, Yang. 2008. Shape Control of Colloidal Metal Nanocrystals. *Small*, 4(3), 310, DOI: 10.1002/smll. 200701295.

Ashkarran, Ali Akbar. 2012. Synthesis and characterization of gold nanoparticles via submerged arc discharge based on a seed-mediated approach. *Journal of Theoretical and Applied Physics.* 6, 14. doi.org/10.1186/2251-7235-6-14.

Belenkov, E.A. and Greshnyakov, V.A. 2015. Diamond-like phases prepared from graphene layers. *Physics of Solid State.* **57**, 205-212.

Berlu, L. and Safouhi, H.J. 2005. Analytical development of multicenter overlap-like quantum similarity integrals over Slater type orbitals and numerical evaluation. *Journal of Theoretical and Computational Chemistry.* 4(3), 787-801.

Berlu, L. and Safouhi, H. and Haggan, P. 2004. Fast and accurate evaluation of three-center, two-electron Coulomb, hybrid, and three-center nuclear attraction integrals over Slater-type orbitals using the SD transformation. *International journal of quantum chemistry.* 99, 221-235.

Castro, Neto A.H., Guinea, F., Peres, N.M.R., Novoselov, K.S., and Geim, A.K. 2009. The electronic properties of graphene. *Reviews of Modern Physics.* 81, 109-162.

Dmitriev, I.S. 1986. *Electron by eyes of chemists.* Leningrad. Chemistry.

Fodorov, A.S., Sorokin, P.B., Avramov, P.V., Ovchinnikov, S.G. 2006. *Modeling of the properties of the electronic structure of some carbon and non-carbon nanoclusters and their interaction with light elements.* Novosibirsk, SO Russian AN.

Gasanov, A.G. and Pashaev, F.G. 2016. The Computer Program for the Study of Nanoparticles in Basis of Slater Atomic Orbitals. *Romanian journal of information science and technology*. 19(4), 331–337.

Gasanov, A.G. and Bayramov, A.A. 2019. Simulation of the Electronic Structure of Graphene–Polyvinylidene Fluoride Composite Material. *Physics of the Solid State*. 61(1), 208–213.

Geim, A.K. and Novoselov, K.S. 2007. The rise of graphene. *Nature Materials*. 6, 182-191.

Gradshteyn, I.S., Ryzhik, I.M. 1980. *Tables of integrals, Sums, Series and Products*, 4th ed. New York. Academic Press.

Guseinov, I.I. 2005. One-range addition theorems for derivatives of integer and noninteger u Coulomb–Yukawa type central and noncentral potentials and their application to multicenter integrals of integer and noninteger n Slater orbitals. *Journal of Molecular Structure: THEOCHEM*. 757, 165-169.

Guseinov, I.I. 2003. Addition theorems for Slater-type orbitals and their application to multicenter multielectron integrals of central and noncentral interaction potentials. *Journal of Molecular Modeling*. 9, 190-194.

Guseinov, I.I., Mursalov, T.M., Paşayev, F.G., Mamedov, B.A., Allahverdiyev, H.A. 1989. Slater tipi orbitaller üzerinden içeren bir ve iki elektronlu çok merkezli integrallerinin hesaplanması. *Journal of Structural Chemistry*. 30,186-188.

Guseinov, I.I. 2003. Comment on "Evaluation of two-center overlap and nuclear-attraction integrals over Slater-type orbitals with integer and noninteger principal quantum numbers". *International Journal of Quantum Chemistry*. 91, 62-64.

Guseinov, I.I. 1998. Analytical evaluation of molecular electric and magnetic multipole moment integrals over Slater-type orbitals, *International Journal of Quantum Chemistry*. 68, 145-150.

Jiang, J.W. and Park, H.S. 2016. Negative Poisson's Ratio in Single-Layer Graphene Ribbons. *Nano Letter.* 16, 2657-2662.

Katsnelson, M.I. 2012. *Graphene: Carbon in Two Dimensions.* New York. Cambridge Univ. Press.

Liu, X., Atwater, M., Wang, J., & Huo, Q. 2007. Extinction coefficient of gold nanoparticles with different sizes and different capping ligands. *Colloids and Surfaces B: Biointerfaces*, 58(1), 3-7.

Maharramov, A.M., Ramazanov, M.A., Gasanov, A.G. and Pashaev, F.G. 2016. The Study of Silver Nanoparticles in Basis of Slater Functions. *Physical Science International Journal.* 10(3), 1-6.

Maharramov, A.M., Aliyeva, T.A., Aliyev, I.A., Pashaev, F.H., Gasanov, A.G., Azimova, S.I., Askerov, R.K., Kurbanov, A.V., Mahmudov, K.T. 2010. Quantum-chemical calculations, tautomeric, thermodynamic, spectroscopic and X-ray studies of 3-(4-fluorophenylhydrazone)pentane-2,4-dione. *Journal of Dyes and Pigments.* 85(1-2), 1-6.

Minkin, V.I., Simkin, B.Y., Minyaev, R.M. 2010. *Theory of structure of molecule.* Rostov at Don, Feniks.

Novoselov, K.S., McCann, E., Morozov, S.V., Fal'ko, V.I., Katsnelson, M.I., Zeitler, U., Jiang, D., Schedin, F., and Geim, A.K. 2006. Unconventional quantum Hall effect and Berry's phase of 2pi in bilayer graphene. *Nature Physics.* 2, 177-180.

Openov, L.A., and Podlivaev, A.I. 2017. Negative Poisson's ratio in a nonplanar phagraphene. *Physics Solid State.* 59, 1267-1269.

Pashaev F.G. 2009. Use of Filter-Steinborn B and Guseinov Qqns auxiliary functions in evaluation of two-center overlap integrals over Slater type orbitals. *Journal of Mathematical Chemistry.* 45, 884–890.

Pashaev, F.G., Gasanov, A.G. and Mahmood, A.T. 2014. The Study of Gold Nanoparticles in basis of Slater Functions. *Journal of Nanotechnology and Advanced Materials.* 2(1), 35-41.

Ramazanov, M.A., Pashaev, F.G., Gasanov, A.G., Maharramov, A., Mahmood, A.T. 2014. The quantum mechanical study of cadmium sulfur nanoparticles in basis of STO's. *Chalcogenide Letters*, 11(7), 359-364.

Ramazanov, M.A., Gasanov, A.G., Pashaev, F.G., and Vahabova, M.R. 2017. Mathematical modeling of PP+(PbS)$_8$+(CdS)$_9$ nanocomposite electron structure by PM3 method. *Fizika*. XXIII, 3, 5-9.

Ranjbartoreh, A.R., Wang, B., Shen, X., and Wang, G. 2011. Advanced mechanical properties of graphene paper. *Journal of Applied Physics*. 109, 014306.

Safouhi, H. 2004. Highly accurate numerical results for three-center nuclear attraction and two-electron Coulomb and exchange integrals over Slater-type functions. *International Journal of Quantum Chemistry*. 100, 172-183.

Shembelov, G.A. 1980. *Quantum chemical methods of calculation of nanoparticles*. Chemistry, Moscow.

Slater, J.C. 1960. *Quantum theory of atomic structure*. Vol. 2. McGraw-Hill, New York.

Stewart, J.J.P. 1989. Optimization of parameters for semiempirical methods I. Method. *Journal of Computational Chemistry*. 10, 209-220.

Zavodinsky, V.G. and Mikhailenko, E.A. 2004. Computer Investigation of Carbon Nanoclusters and Their Activities in Reactions with Molecular Oxygen. *Physics of. Low-Dimensional Structure*. 5–6, 35-48.

Zavodinsky, V.G., Kuyanov, I.A., and Chukurov, E.N. 1999. Energetics of boron in the Si(111)-($\sqrt{3}\times\sqrt{3}$)-B surface phase and in subsurface silicon layers. *Surface Review and Letters*. 6, 127-132.

Zavodinskii, V.G., Chibisov, A.N., Gnidenko, A.A., and Aleinikova, M.A. 2005. Theoretical investigation of elastic properties of small nanoparticles with various types interatomic bonds. *Mekhanika Kompozitionnix Materialov i Konstruktsiy*. 11, 337-346.

In: An Introduction to Electronic Structure ... ISBN: 978-1-53618-411-2
Editor: Nadia T. Paulsen © 2020 Nova Science Publishers, Inc.

Chapter 4

BASICS AND APPLICATIONS OF ELECTRONIC STRUCTURE THEORY

Aditya M. Vora[*]

Department of Physics, University School of Sciences, Gujarat University, Navrangpura, Ahmedabad, Gujarat, India

ABSTRACT

Recently, the electronic structure theory plays an important role for studying the various physical properties of materials through ab-initio approach. It depends equally upon density functional theory (DFT) and wave function theory (WFT). The purpose of this chapter is to give a basic introduction of electronic structure theory with commonly used notation and its applications for studying the physical properties of materials.

Keywords: basics of electronic structure theory, applications of electronic structure theory, wave function theory; density functional theory.

---

[*] Corresponding Author's Email: voraam@gmail.com.

INTRODUCTION

The progress of quantum mechanics is debatably one of the utmost advances in physics in the last era. With the advent of quantum mechanics, we expanded precious vision into the nature of the atom and can now dependably model and envisage the interactions of subatomic particles. The computational physics and chemistry apply the concepts of quantum mechanics to the higher scale area of molecules and reactions between them. The level of control afforded by computational methods permits one to probe queries which may not be conceivable to answer via experimental methods. The computational physics and chemistry consent us to achieve numerous things that are not yet possible experimentally such as following the real time dynamics of an electron as it transfers from its ground state energy level up through several excited states and back again to the ground state.

The work on the study of the electronic structure of materials is at a significant phase, with new algorithms and computational procedures and fast developments in basic theory. Many properties of materials can now be determined directly from the fundamental equations for the electrons, p roviding new insights into the critical problems in physics, chemistry and materials science [1-99].

The goal of electronic structure theory is to computationally calculate the electronic Schrödinger equation with some modification by the special theory of relativity, if essential to recognize and even envisage the properties and alterations of chemical systems. Meanwhile, the equation is the exact equation of motion of essential particles, the electronic structure theory can in principle deliver whole quantitative facts of these properties and transformations [71], rotating a progressively broader area of chemistry into the computational science. This may appear to be an exaggeration, but for minor gas-phase molecules made of light nuclei, the electronic structure theory has definitely started to envisage a change of their properties, e.g., shape and color with such high accurateness [72] that it can compete with experimental methods and change the traditions in which research is led in areas like combustion and interstellar chemistry.

Basics and Applications of Electronic Structure Theory

The electronic structure theorists are determined to carry such changes in all areas of chemistry by growing the accuracy, efficiency and applicability of their computational approaches [70].

Electronic structure theory describes the motions of electrons in atoms or molecules. Generally, this is done in the context of the Born-Oppenheimer approximation, which says that electrons are so much lighter (and therefore faster) than nuclei that they will find their optimal distribution for any given nuclear configuration. The electronic energy at each nuclear configuration is the potential energy that the nuclei feel, so solving the electronic problem for a range of nuclear configurations gives the potential energy surface [1-72].

Since the electrons are consequently small, one wants to practice quantum mechanics to explain for their motion. The quantum mechanics expresses us that the electrons will not be localized at particular points in space, but they are best assumed of as 'matter waves' which can interfere. The probability of finding a single electron at a specified point in space is represented by $\Psi^*(x)\Psi(x)$ for its wavefunction Ψ at the point x and it is determined by solving the time-independent Schrödinger equation $\hat{H}\Psi = E\Psi$. If the problem is related to time-dependent, then the time-dependent Schrödinger equation $\hat{H}\Psi = i\hbar \frac{\partial \Psi}{\partial t}$ must be used instead; else, the solutions to the time-independent case are also answers to the time-dependent problem when they are multiplied by the energy dependent phase factor $e^{-iEt/\hbar}$. Meanwhile we have immobile the nuclei under the Born-Oppenheimer approximation, we explain for the non-relativistic electronic Schrödinger equation as follows [1-72]:

$$\hat{H} = -\frac{\hbar^2}{2m}\Sigma_l \nabla_l^2 - \Sigma_C \frac{\hbar^2}{2M_C}\nabla_C^2 - \Sigma_{C,l}\frac{Z_C e^2}{4\pi\epsilon_0 r_{Cl}} + \Sigma_{C>D}\frac{Z_C Z_D e^2}{4\pi\epsilon_0 R_{CD}} + \Sigma_{l>m}\frac{e^2}{4\pi\epsilon_0 r_{lm}}. \qquad (1)$$

Where, l, m mention to electrons and C, D refer to nuclei. In atomic units, it simplifies to

$$\hat{H} = -\frac{1}{2}\Sigma_l \nabla_l^2 - \Sigma_C \frac{1}{2M_C}\nabla_C^2 - \Sigma_{C,l}\frac{Z_C}{r_{Cl}} + \Sigma_{C>D}\frac{Z_C Z_D}{R_{CD}} + \Sigma_{l>m}\frac{1}{r_{lm}}. \quad (2)$$

Such Hamiltonian is suitable as long as for relativistic properties, which are not significant for the system in question. The relativistic effects are not usually considered vital for atoms with atomic number (Z) below about 25 (Mn). For heavier atoms, the internal electrons are held more tightly to the nucleus and have velocities which increase as the atomic number (Z) of elements increases; as these velocities approach the speed of light, the relativistic effects become more significant. However, there are many methods for accounting for the relativistic effects, but the most general is to use relativistic effective core potentials (RECPs), frequently along with the typical non-relativistic Hamiltonian above [70].

One must instead view this as a chemistry or physics problem and remove many of these difficulties, e.g., the high dimensionality, singularities and complex dependencies due to anti-symmetry by clarifying and exploiting the structures of electronic wavefunctions. Let us consider the singularities in the Hamiltonian due to electron-nucleus combination. The occurrence of these singularities means that the electrons sufficiently close a nucleus in a molecule experience only the nucleus' strong attractive forces and therefore mainly behave like atomic electrons. It advises the usage of atomic-orbital (AO) basis functions and linear-combination-of-atomic-orbital (LCAO) expansions of molecular spin-orbitals (MO) as a rational explanation of molecular wavefunctions. The electron-electron interactions also cause numerous singularities in the Hamiltonian, but they are repulsive. It means that the electrons try to be as far left as possible from one another and their motions are, therefore, only feebly correlated. This recommends a calculation of a wavefunction by an anti-symmetrized product by Slater determinant of LCAO MOs, each of which accommodates one electron. This product form of the wavefunction presents an approximate parting of variables, extremely dropping the effective dimension of the equation of motion. The Hartree-Fock (HF) method is defined as such and constantly recovers ca. 99% of error for a wider range of chemical systems. However, the HF method accomplishes extraordinary correctness at a small computational cost, the accuracy is far from suitable for quantitative chemistry. One must improve the so-called electron-correlation energy, which is defined as the alteration between the HF and exact energies [70].

The time related theories are based on diagrammatic many-body theories that are size extensive in the sense that they keep even precision across many system sizes and are valid similarly to problems in quantum chemistry as well as in nuclear, atomic, solid state and condensed matter physics. The many-body perturbation theory (MBPT) and coupled cluster (CC) methods [73] are perhaps the most successful examples in such category and are used mostly in quantifiable chemical simulations currently. This is in difference with the configuration interaction (CI) method, which is neither an illustrative many-body method nor size extensive and hence gradually going from the important electron-correlation arsenals. Despite the considerable development in the field, challenges continue. In the next are listed some of the most significant research problems nowadays, some nearly fixed and others still being in a former phase of examination. The cutting boundaries of research addressing various features of the electronic structure theory are discussed very briefly by researchers [74-76].

To solve the Schrödinger equation, we have to find appropriate approximations. Various outlines lead to different approach that lead themselves to dissimilar approximations, that in turn work for different systems and circumstances. However, the wavefunction $\Psi(x,t)$ are based on following methods [1-99].

Density $\rho(r)$ Based Methods, i.e., Density Functional Theory (DFT)

Such theory reformulates the problem in terms of the electronic density and can be applied to finite and periodic systems. A wide variety of different functionals exist the task with such methods is resulting good and systematically improvable approximations for the functional.

The DFT does not obey to the outline of wavefunction theory (WFT) viz. the one founded on the development of an exact wavefunction by determinants. It instead objects at improving the precise energy as a functional of electron density. Eliminating at least partially the high-dimensional wavefunction as an extent to be determined, the DFT inclines to accomplish greater cost correctness performance and better practical usefulness than the present complements in WFT. While neither regular nor analytical in the sense of CC, MBPT or RPA, approximations in DFT have seen an explosive advance in the past two eras, removing such limitations in previous approximations as their poor descriptions of ionization energies [77], Rydberg excitation energies [77], charge transfer states [78], (hyper)polarizabilities [79], dispersion [80] and stacking interactions [81]. The usage of electron density in lieu of a wavefunctions as a basic variable essential, thus, somehow exploit an unseen assembly of wavefunctions, the feature of which has not been unstated. Additionally, the theoretical studies of vital nature are consequently still acceptable to explain why some approximations of the DFT achieve so well for a range of difficulties. Eventually, like the WFT and the DFT wants to be considered by systematic sequences of approximations converging toward the accuracy, in which electron density may be a low-rank member of a new order of fundamental physical variables such as density matrices [82], intracules and two-electron density parameters etc. In which, the last two have been supported by Gill and coworkers [76, 83].

Explicitly Correlated Methods

The weak electron correlation is taken by CC and MBPT that define equally avoiding motion of electrons as a superposition of excited determinants. It is a difficult physical image of correlation, which has a serious practical problem; the convergence of correlation energies hence got is very slow with respect to the size of the AO basis set. A more physically attractive and efficient method is to present a basis function that depends clearly on interelectronic distances (R_{12}), which can define points in wavefunctions that happen because of the electron-electron singularities in the Hamiltonian [84]. The last era has seen extraordinary development in such class of approaches, e.g., $R12$ method [85], virtually eliminating the problem of slow basis-set convergence. Two of the advances behind such advancement are the finding [86] of a nearly optimum form of the R_{12}-dependent basis function or so-called Slater-type correlation factor, which is introducing the $F12$ method and the application [87] to determine the development coefficient multiplying the R_{12} dependent basis function analytically as suggested by Ten-no [74].

Condensed-Phase and Complex Systems

Efficient, size-extensive electron-correlation approximations like as CC, MBPT and their combinations converging near to the precise basis-set explanations of the Schrödinger equation have been established increasingly with the support of computer algebra [88]. In which, regular basis sets converging toward comprehensiveness are also available. They have been involved in the construction of the electronic structure theory, an analytical science at tiniest for small molecules in the gas phase. Currently, these methods also advance the view of having condensed-phase systems and large composite systems such as biological systems, in which the strengths of the relevant interactions, including dispersion and stacking interactions [89], extent three orders of magnitude in the appropriate area of predictive simulations, potentially altering quantitative features of these fields. Significant development has already been completed to the use of spatially local basis functions that exploits the fundamentally local nature of correlation and speeds up these computations dramatically [90]. In addition to the speedup actions that benefit all these methods, an illustrative many-body process that strikes higher balance between correctness and computational price and is subject to an extension to metallic solids or to an addition with statistical thermodynamics, e.g., a finite-temperature extension, continues to be reasonable. The random-phase approximation (RPA), familiarized to chemistry by Furche [91, 92], is one such process which is only slightly more expensive than HF yet is proficient of taking much of electron correlation. The RPA, like CC [93], contains the sum of perturbation alterations of a certain type to an infinite order and can therefore resist a breakdown characteristic of MBPT when quasi-degenerate states are presented close the ground state such as arise in metals. Its possibilities to be a practical, systematic and size-extensive electron-correlation technique suitable for small and large molecules as well as solids similarly. Also, such methods trust on sequentially refining the correctness of the approximated wavefunction in a classified way. Therefore, these methods are mostly useful to finite systems, e.g., molecules and clusters. Usually used methods from this class are as follows Hartree-Fock (HF), Mjøller-Plesset (MP) perturbation theory and Coupled Cluster (CC). Many body perturbation theory-based methods are frequently used in solid state physics and quantum chemistry and it can be thoroughly improved. Some well-known methods of group are GW, T-Matrix, BSE and FLEX, respectively [70].

Strong Correlation

Strongly correlated systems are ones in which there are two or more quasi-degenerate states are interrelating with one another. They contain such arrangements as spin lattices, transition-metal complexes with many metal cores and breaking and creation of chemical bonds. The HF method, in which electron correlation is supposed indirectly to be weak, becomes a poor approximation, rendering MBPT and rarely even CC built on a single reference determinant also insufficient. There has been a remarkable collection of unique methods newly proposed to resolve such problem, e.g., including various multi-reference and active-space electron-correlation methods, selectively sum critical determinant contributions inside the context of WFT [94]. While, some of them involves the number of electrons to be varied [95]. Conceivably the most fruitful ones hence far basically option to the full CI method but with massively better algorithms based on quantum Monte Carlo (QMC) [96], density matrix renormalization group [97], graphically contracted functions [98], etc. These methods and algorithms indicate that certain regular structures occur in strongly correlated wavefunctions, which are yet to be completely understood. The field of electronic structure theory has experienced sharp and linear progress since 1990, capitalizing upon the introductory work in the previous years, into a new level of maturity. It has altered neighboring experimental fields of chemistry and physics, which, as a result, rely progressively deeply on black-box computational methods and software based on electronic structure theorists have established. It also remains to attend as an important infrastructure of advanced quality and superior suitability for the associated fields of theoretical chemistry and physics that is quantum dynamics and statistical thermodynamics. There is hardly any question that this tendency will remain and the main of electronic structure theory is still gaining. The QMC methods are founded on a stochastic result of the Schrödinger equation and representation of $\Psi(x,t)$ and it can be applied to finite as well as periodic systems. Common variations are Variational and Diffusion Monte Carlo methods. The density matrix functional theory is one of less explored methods. The vital object about the different outlines is that all of them are in principle exact. Subsequently, they are based on different quantities and make different approximations, they provide us a miscellaneous set of tools to study a system [70].

DEVELOPMENT OF ELECTRONIC STRUCTURE THEORY

The 3D periodic solids are amongst the first systems for which the electronic structure theory is developed. The notion of an electronic band structure scopes back to the first era after the development of quantum mechanics and early illustrations can be found [56, 57]. Nearly fifty years before, the total energy of the electrons and nuclei in a basic unit cell of a crystalline solid has originate into focus of theoretical researches. To find it, summation over reciprocal space or k-space is essential. Therefore, through the 1970s, a number of research papers about the specimen of k-

space have been published [6, 13, 15, 28, 42]. Assumed that computational power in these days was somewhat inadequate, the importance at this period was to retain the number of k-points to be preserved as small as conceivable and the most efficient selection of special k-points was the major issue of these previous works [67].

With the rise of computer power, the computational physicists and materials scientists have started their work on additional and more complex systems including hundreds of atoms in one-unit cell. Meanwhile, this large unit cell goes along with minor Brillouin zone (BZ) in reciprocal space, specimen of k-space established less consideration. Furthermore, liquid or amorphous systems that lack translational order are approximated by large supercells, e.g., using quasi-random structures. Due to the nonexistence of real physical periodicity, computations with big supercells frequently pay fair one k-point which, for the sake of further computational reserves, is repeatedly selected to be the $\hat{W}$-point, i.e., the source in reciprocal space. Recently, the demand to do highly precise calculations has raised renewed attention in better methods for the sampling of reciprocal space.

One driving factor initiates from computational materials science. For thermodynamic studies, e.g., for the computations of phase diagrams, highly converged total energies for the basic unit cells of bulk materials are mandatory [22, 23]. The essential intentions should be achieved in an automated mode, using the methods of high-throughput computing. Therefore, one uses automatically produced, very condensed k-point sets that permit one to influence an exactness of the total energy better than 1 meV per atom. As has been exposed in a recent study of Morgan et al. [44], in order to assure such precision level for all phases with differently sized and shaped unit cells, a k-point density as high as 5,000 k-points/Å-3 is classically essential. Also, the methods based on machine learning effort to select k-point grids that are utmost appropriate for the problem at the hand [14]. As another factor driving invention in the field of k-point sampling, the attention in special properties of bulk materials, in specific in the areas of electronic transport, magnetism and topological states of matter, has led to enhanced, e.g., adaptive structures. Though, the total energy is variational with respect to small changes in the charge density and is hence computationally strong, the applications stated above need the determination of very fine structures in the Brillouin zone in order to get a correct explanation of the properties of attention.

Very recently, the DFT has fascinated strong attention and significant development has been attained [11, 24], which describes the wavefunction in terms of single-particle orbitals. Efficient computer codes for Hartree-Fock (HF) computations including the selection to treat period systems [16] are available in the literature so far. Here, the difficulty of treating correlations between two electrons in dissimilar unit cells of the solid must be undertaken with. The finest associated ground-state wavefunction in a variational sense may have less symmetries than the many-particle Hamiltonian; thus, approaches exploiting the translational symmetry of the crystal want to be measured with caution. A systematic way to include electronic correlations in calculations for periodic systems is offered by the way of additions [48]. Additional choice appropriate to the periodic systems is the Quantum Monte Carlo (QMC) method suggested by Foulkes et al. [18] which permits for an even more flexible mathematical representation of the numerous particle wavefunction than they are starting from a basis set of atomic orbitals.

PROPERTIES PROJECTED BY ELECTRONIC STRUCTURE THEORY

According to one of the hypothesis of quantum mechanics, if we recognize the wavefunction $\Psi(x,t)$ for a specified system, then we can determine any property of that system, at least in belief. Apparently, if we use approximations in determining the wavefunction, then the properties found from it will also be approximated [68].

Meanwhile we almost constantly invoke the Born-Oppenheimer approximation, we only have the electronic wavefunction which is not the full wavefunction for electrons and nuclei. Hence, some properties concerning nuclear motion are not essentially accessible in the context of electronic structure theory [68].

To fully recognize the facts of a chemical reaction, we want to use the electronic structure outcomes to carry out following dynamics computations. Luckily, though, fairly a few properties are inside the reach of just the electronic problem. For example, meanwhile the electronic energy is the potential energy felt by the nuclei, minimizing the electronic energy with respect to nuclear coordinates gives an equilibrium formation of the molecule [68].

The electronic wavefunction or its various derivatives are adequate to calculate the following properties [68]:

- Geometrical structures, e.g., rotational spectra,
- Rovibrational energy levels, e.g., infrared and Raman spectra,
- Electronic energy levels, i.e., UV and visible spectra,
- Quantum Mechanics + Statistical Mechanics = Thermochemistry properties such as $\Delta H, \Delta S, \Delta G, C_V$ and C_P and primarily gas phase,
- Ionization potentials, e.g., photoelectron and X-ray spectra,
- Potential energy surfaces, i.e., barrier heights and transition states with an action of dynamics and this leads to reaction rates and mechanisms,
- Franck-Condon factors, e.g., transition probabilities and vibronic intensities,
- Electron affinities,
- Dipole moments,
- IR and Raman intensities,
- Polarizabilities,
- Electron density maps and population analyses, and
- Magnetic shielding tensors, e.g., NMR spectra.

The main goal and role of computational electronic structure theory is displayed in Figure 1.

Basics and Applications of Electronic Structure Theory   143

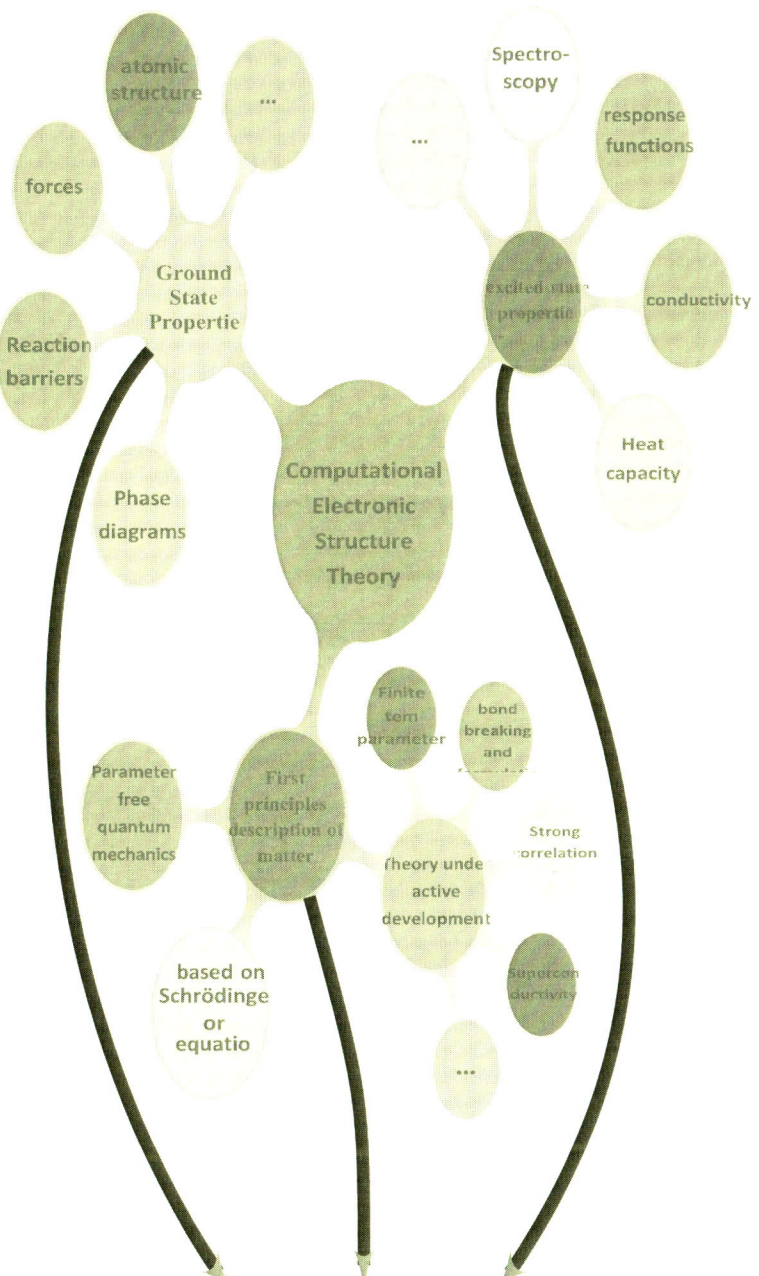

Figure 1. The goal and role of computational electronic structure theory [68].

RESULTS AND DISCUSSION

In present article, an attempt has been made to perform a detailed and comparable study of some basics and applications of the electronic structure theory. With the cumulative part of large amount of data and higher throughput computations in materials science particularly physics and chemistry, it is significant to promise high exactness for the first-principles information to be kept in materials records. Because of this, there is newly attracted in convergence facets of the DFT based computations. There is a sufficient information on the convergence with the sampling of the reciprocal space from previous effort with considerable computational resources. For semiconductors and insulators, such information about special k-points can be used to take away the calculations in the most economical way. For metals and alloys, the exact specimen of the Fermi surface is still a subject and needs the use of dense k-point meshes to confirm a convergence of the total energy per atom to better than 1meV. Even for simple metals, there is surprising variety in the forms of their Fermi surfaces. Therefore, there is currently no universal procedure accessible for picking k-points that avoids a dense sampling of the Brillouin zone. Besides, for high sampling density, the exact place of each k-point, and so the information about special k-point groups, becomes fewer appropriate. If the goal of the computations is the total energy and the relaxed atomic structure, then the utmost efficient mode to deal with the Fermi surface is approximately enlargement of the Fermi-Dirac (FD) distribution function. The expansion must be selected such that a sufficiently huge segment of all occupation figures displays fractional occupation. The approaches obtainable for generalizing the total energy to the limit of zero enlargement work well also with a very high number of k-points. The forces on the atoms warrant special consideration, subsequently they are calculated as the derivative of the electronic free energy somewhat the total energy, then for the test cases examined this did not cause difficulties as long as with a broadening parameter in the range of 0.1-0.2 eV and a sufficiently dense k-point mesh is utilized. For post-processing with investigation tools and for extends outside the total energy, e.g., for transport properties, the addition schemes that go beyond the typical trapezoidal rule, e.g., quadratic interpolation or adaptive k-point meshes, turn out to be valuable and offer an area for upcoming research [67].

Conclusion

In the present article, some basics and applications of electronic structure theory are discussed briefly. Such chapter will find very much helpful to the scientific community working in the field of electronic structure theory.

Acknowledgments

The author is highly acknowledged for the computer facility developed under DST-FIST programme from Department of Science and Technology, Government of India, New Delhi, India and financial assistance under DRS-SAP-II from University Grants commission, New Delhi, India.

References

[1] Ashcroft, N. W., and Mermin, N. D. (1976). *Solid State Physics.* Philadelphia, PA: Saunders College.

[2] Ziman, J. M. (1972). *Principles of the theory of solids.* 2nd Edition. Cambridge: University Press.

[3] Szabo, A., and Ostlund. N. S. (1996). *Modern Quantum Chemistry: Introduction to Advanced Electronic Structure Theory.* Mineola, NY: Dover Publications.

[4] Szabo A., and Ostlund, N. S. (1989). *Modern Quantum Chemistry: Introduction to Advanced Electronic Structure Theory*. NY: McGraw-Hill.

[5] Perdew, J., and Zunger, A. (1981) Self-interaction correction to density-functional approximations for many-electron systems. *Phys. Rev.* B23: 5048-5079.

[6] Baldereschi, A. (1973). Mean-value point in the Brillouin zone. *Phys. Rev.* B7:5212-5215.

[7] Bloch, F. (1928). Über die Quantenmechanik der Elektronen in Kristallgittern [About the quantum mechanics of electrons in crystal lattices]. *Z. Physik* 52:555-600.

[8] Blöchl, P. (1996). Projector augmented-wave method. *Phys. Rev.* B50:17953-17979.

[9] Blöchl, P. E., Jepsen, O., and Andersen, O. K. (1994). Improved tetrahedron methods for Brillion zone integration. *Phys. Rev.* B49:16223-16233.

[10] Blum, V., Gehrke, R., Hanke, F., Havu, P., Havu, V., Ren, X., et al. (2009). Ab initio molecular simulations with numeric atom-centered orbitals: FHI-aims. *Comp. Phys. Commun.* 180:2175-2196.

[11] Booth, G. H., Grüneis, A., Kresse, G., and Alavi, A. (2012). Towards an exact description of electronic wavefunctions in real solids. *Nature* 493:365-370.

[12] Bruno, E., and Ginatempo, B. (1997). Algorithms for Korringa-Kohn-Rostoker electronic structure calculations in any Bravais lattice. *Phys. Rev.* B55:12946-129552.

[13] Chadi, D. J., and Cohen, M. L. (1973). Special points in the brillouin zone. *Phys. Rev.* B 8:5747-5753.

[14] Choudhary, K., and Tavazza, F. (2019). Convergence and machine learning predictions of Monlhorst-Pack k-point and plane-wave cut-off in high-throughput DFT calculations. *Comput. Mater. Sci.* 161:300-308.

[15] Cunningham, S. L. (1974). Special points in the two-dimensional Brillouin zone. *Phys. Rev.* B10:4988-4994.

[16] Dovesi, R., Erba, A., Orlando, R., Zicovich-Wilson, C. M., Civalleri, B., Maschio, L., et al. (2018). Quantum-mechanical condensed matter simulations with crystal. Wiley Interdiscip. *Rev. Comput. Mol. Sci.* 8:e1360.

[17] Erwin, S. C., Barke, I., and Himpsel, F. J. (2009). Structure and energetics of Si (111) - (5×2) - Au. *Phys. Rev.* B80:155409 (1-10).

[18] Foulkes, W. M. C., Mitas, L., Needs, R. J., and Rajagopal, G. (2001). Quantum Monte Carlo simulations of solids. *Rev. Mod. Phys.* 73: 33-83.

[19] Freysoldt, C., Boeck, S., and Neugebauer, J. (2009). Direct minimization technique for metals in density functional theory. *Phys. Rev.* B79:241103 (1-4).

[20] Froyen, S. (1989). Brilloiun zone integration by Fourier quadrature: special points for superlattice and supercell calculations. *Phys. Rev.* B39:3168-3172.

[21] Gillan, M. J. (1989). Calculation of the vacancy formation energy in aluminium. *J. Phys.* 1:689-711.

[22] Grabowksi, B., Hickel, T., and Neugebauer, J. (2007). Ab initio study of the thermodynamic properties of nonmagnetic elementary fcc metals: exchange correlation-related error bars and chemical trends. *Phys. Rev.* B76:024309 (1-16).

[23] Grabowski, B., Söderlind, P., Hickel, T., and Neugebauer, J. (2011). Temperature driven phase transitions from first principles including all relevant excitations: The fcc-to-bcc transition in Ca. *Phys. Rev.* B84:214107 (1-20).

[24] Gruber, T., Liao, K., Tsatsoulis, T., Hummel, F., and Grüneis, A. (2018). Applying the coupled-cluster ansatz to solids and surfaces in the thermodynamic limit. *Phys. Rev.* X 8:021043 (1-28).

[25] Hart, G. L. W., Jorgensen, J. J., Morgan, W. S., and Forcade, R. W. (2018). *A robust algorithm for k-point grid generation and symmetry reduction.* ArXiv:1809:10261v2.

[26] Heiliger, C., Zahn, P., and Yavorsky, Y., and Mertig, I. (2008). Thickness dependence of the tunnelling current in the coherent limit of transport. *Phys. Rev.* B77:224407 (1-6).

[27] Henk, J. (2001). Integration over two-dimensional Brillouin zones by adaptive mesh refinement. *Phys. Rev.* B64:035412 (1-6).

[28] Jepsen, O., and Andersen, O. K. (1971). Linear tetrahedron methods for Brillouin zone integration. *Solid State Commun.* 9:1763 (1-8).

[29] Kahnouji, H., Kratzer, P., and Hashemifar, S. J. (2019). Ab initio simulation of the structure and transport properties of zirconium and ferromagnetic cobalt contacts on the two-dimensional semiconductor WS_2. *Phys. Rev.* B99:035418 (1-9).

[30] Kaminski, J. W., Kratzer, P., and Ratsch, C. (2017). Towards a standardized setup for surface energy calculations. *Phys. Rev.* B95:085408 (1-11).

[31] Kiejna, A., Peisert, J., and Scharoch, P. (1999). Quantum-Size effect in thin Al (110) slabs. *Surf. Sci.* 432:54-60.

[32] Koch, W., and Holthausen, M. C. (2001). *A Chemist's Guide to Density Functional Theory.* Weinheim: Wiley-VCH.

[33] Kresse, G., and Furthmüller, J. (1996a). Efficiency of ab initio total energy calculations for metals and semiconductors using a plane-wave basis set. *Comp. Mater. Sci.* 6:15-50.

[34] Kresse, G., and Furthmüller, J.(1996b) Efficient iterative schemes for a binitiototal energy calculations using a plane-wave basis set. *Phys. Rev.* B54:11169-11186.

[35] Kresse, G., and Joubert, D. (1999). From ultrasoft pseudopotentials to the projector augmented-wave method. *Phys. Rev.* B59:1758-1775.

[36] MacDonald, A. H., Vosko, S. H., and Coleridge, P. T. (1979). Extensions of the tetrahedron method for evaluating spectral properties of solids. *J. Phys. C Solid State Phys.* 12:2991-3002.

[37] Martin, R. M. (2004). *Electronic Structure-Basic Theory and Practical Methods,* Cambridge, UK: Cambridge University Press.

[38] Marzari, N., Vanderbilt, D., De Vita, A., and Payne, M. C. (1999). Thermal contraction and disordering of the Al(110) surface. *Phys. Rev. Lett.* 82:3296-3299.

[39] Marzari, N., Vanderbilt, D., and Payne, M. C. (1997). Ensemble density-functional theory for ab initio molecular dynamics of metals and finite-temperature insulators. *Phys. Rev. Lett.* 79:1337-1340.

[40] Mermin, N. D. (1965). Thermal properties of the inhomogeneous electron gas. *Phys. Rev.* 137:A1441-A1443.

[41] Methfessel, M., and Paxton, A. T. (1989). High-precision sampling for Brillouin-zone integration in metals. *Phys. Rev.* B40: 3616-3621.

[42] Monkhorst, H. J., and Pack, J. D. (1976). Special points for Brillouin-zone integrations. *Phys. Rev.* B13:5188-5192.

[43] Moreno, J., and Soler, J. M. (1992). Optimal meshes for integrals in real-and reciprocal-space unit cells. *Phys. Rev.* B45:13891-13898.

[44] Morgan, W. S., Jorgensen, J. J., Hess, B. C., and Hart, G. L. W. (2018). Efficiency of generalized regular k-point grids. *Comp. Mater. Sci.* 153:424-430.

[45] Neugebauer, J., and Scheffler, M. (1992). Adsorbate-substrate and adsorb ate adsorbate interactions of Na and K adlayers on Al(111). *Phys. Rev.* B46:16067-16080.

[46] Pack, J. D., and Monkhorst, H. J. (1977). Special points for Brillouin-zone integrations-a reply. *Phys. Rev.* B16:1748-1749.

[47] Parr, R. G., and Weitao, Y. (1994). Density-Functional Theory of Atoms and Molecules. *International Series of Monographs on Chemistry*, Oxford: Oxford University Press.

[48] Paulus, B. (2006). The method of increments-a wavefunction-based ab initio correlation method for solids. *Phys. Rep.* 428:1-52.

[49] Popescu, V., and Kratzer, P. (2013). Large Seebeck magnetic anisotropy in thin Co films embedded in Cu determined by ab initio investigations. *Phys. Rev.* B88:104425 (1-7).

[50] Rajagopal, G., Needs, R. J., James, A., Kenny, S. D., and Foulkes, W. M. C. (1995). Variational and diffusion Monte Carlo calculations at nonzero wave vector: Theory and application to diamond-structure germanium. *Rev. Phys.* B51:10591-10600.

[51] Ramstad, A., Brocks, G., and Kelly, P. J. (1995). Theoretical Study of the Si(100) surface reconstruction. *Phys. Rev.* B51:14504-14523.

[52] Razee, S. S. A., Staunton, J. B., and Pinski, F. J. (1999). First-principles theory of magnetocrystalline anisotropy of disordered alloys: application to cobalt platinum. *Phys. Rev.* B56:8082-8090.

[53] Rignanese, G. M., Ghosez, P., Charlier, J. C., Michenaud, J. P., and Gonze, X. (1995). Scaling hypothesis for corrections to total energy and stress in plane-wave-based ab initio calculations. *Phys. Rev.* B52:8160-8178.

[54] Sholl, D. S., and Steckel, J. A. (2009). *Density Functional Theory-a Practical Introduction.* Hoboken, NJ: Wiley.

[55] Slater, J. C. (1934). The electronic structure of metals. *Rev. Mod. Phys.* 6:209-280.

[56] Sommerfeld, A., and Bethe, H. (1933). Elektronen theorie der Metalle. *Handbuch der Physik,* Vol. 24-2. 2 Edn. Springer, 333.

[57] Spitaler, J., and Estreicher, S. K. (2018). Perspectives on the theory of defects. *Front. Mater.* 5:70 (1-17).

[58] Streitwolf, H. W. (1971). *Group Theory in Solid State Physics.* University Physics Series, Macdonald and Company.

[59] Temmerman, W., and Szotek, Z. (1987). Calculating the electronic structure of random alloys with the KKR-CPA method. *Comp. Phys. Rep.* 55:173-220.

[60] Tersoff, J., and Hamann, D. R. (1983). Theory and application for the scanning tunnelling microscope. *Phys. Rev. Lett.* 50:1998-2001.

[61] van Hove, L. (1953). The occurrence of singularities in the elastic frequency distribution of a crystal. *Phys. Rev.* 89:1189-1193.

[62] Wagner, F., Laloyaux, T., and Scheffler, M. (1998). Errors in Hellmann-Feynman forces due to occupation-number broadening and how they can be corrected. *Phys. Rev.* B57:2102-2107.

[63] Wei, C. M., and Chou, M. Y. (2002). Theory of quantum size effects in thin Pb (111) films. *Phys. Rev.* B66:233408 (1-4).

[64] Wisesa, P., McGill, K., and Mueller, T. (2016). Efficient generation of generalized Monkhorst-Pack grids through the use of informatics. *Phys. Rev.* B93:155109 (1-10).

[65] Zhang, X., Grabowski, B., Körmann, F., Freysoldt, C., and Neugebauer, J. (2017). Accurate electronic free energies of the 3d,

4d, and 5d transition metals at high temperatures. *Phys. Rev. B* 95:165126 (1-13).

[66] https://en.wikipedia.org/wiki/Electronic_structure.

[67] Kartzer, P., and Neugebauer J. (2019). The basics of electronic structure theory for periodic systems, *Front. Chem.* 7:106 (1-18).

[68] Patrick, R. (2014). *A Lecturer note on 'Electronic Structure Theory'*, Fritz Haber Institute of the Max Planck Society Faraday Weg 4-6 Berlin.

[69] Rodney J. B. (2010). Ab initio DFT and its role in electronic structure theory. *Mol. Phys.* Vol. 108: 3299-3311.

[70] Hirata, S. (2012). Electronic structure theory: present and future challenges. *Theor. Chem. Acc.* 131:1071 (1-4).

[71] Schaefer, H. F. III. (1986). Methylene: A Paradigm for Computational Quantum Chemistry. *Science* 231:1100-1107.

[72] Bytautas, L., and Ruedenberg, K. (2006). Correlation energy extrapolation by intrinsic scaling. V. Electronic energy, atomization energy, and enthalpy of formation of water. *J. Chem. Phys.* 124:174304 (1-13).

[73] Bartlett, R. J., and Musiał, M. (2007). Coupled-cluster theory in quantum chemistry. *Rev Mod Phys* 79:291-351.

[74] Ten-no, S. (2012). Explicitly correlated wave functions: summary and perspective. *Theor. Chem. Acc.* 131:1070 (1-11).

[75] Eshuis, H., Bates, J. E., and Furche, F. (2012). Electron correlation methods based on the random phase approximation. *Theor. Chem. Acc.* 131:1084 (1-18)-.

[76] Gill, P. M. W., and Loos, P. -F. (2012). Uniform electron gases. *Theor. Chem. Acc.* 131:1069 (1-9).

[77] Tozer, D. J., and Handy, N. C. (1998). Improving virtual Kohn–Sham orbitals and eigenvalues: Application to excitation energies and static polarizabilities. *J. Chem. Phys.* 109:10180-10189.

[78] Iikura, H., Tsuneda, T., Yanai, T., Hirao, K. (2001). A long-range correction scheme for generalized-gradient-approximation exchange functionals. *J. Chem. Phys.* 115:3540-3544.

[79] Kamiya, M., Sekino, H., Tsuneda, T., Hirao, K. (2005). Nonlinear optical property calculations by the long-range-corrected coupled-perturbed Kohn–Sham method. *J. Chem. Phys.* 122:234111 (1-10).

[80] Kamiya, M., Tsuneda, T., and Hirao, K. (2002). A density functional study of van der Waals interactions. *J. Chem. Phys.* 117:6010-6015.

[81] Zhao, Y., and Truhlar, D. G. (2005). How well can new-generation density functional methods describe stacking interactions in biological systems?. *Phys. Chem. Phys.* 7:2701-2705.

[82] Nakatsuji, H., Yasuda, K. (1996). Direct Determination of the Quantum-Mechanical Density Matrix Using the Density Equation. *Phys. Rev. Lett.* 76:1039-1042.

[83] Gill, P. M. W., O'Neill, D. P., and Besley, N. A. (2003). Two-electron distribution functions and intracules. *Theor. Chem. Acc.* 109:241-250.

[84] Klopper, W., Manby, F. R., Ten-no, S., and Valeev, E. F. (2006). R_{12} methods in explicitly correlated molecular electronic structure theory. *Int. Rev. Phys. Chem.* 25:427-468.

[85] Kutzelnigg, W. (1985). r_{12}-Dependent terms in the wave function as closed sums of partial wave amplitudes for large *l*. *Theor. Chim. Acta* 68:445-469.

[86] Ten-no, S. (2004). Initiation of explicitly correlated Slater-type geminal theory. *Chem. Phys. Lett.* 398:56-61.

[87] Ten-no, S. (2004). Explicitly correlated second order perturbation theory: Introduction of a rational generator and numerical quadratures. *J. Chem. Phys.* 121:117-128.

[88] Hirata, S. (2006). Symbolic Algebra in Quantum Chemistry. *Theor. Chem. Acc.* 116:2-17.

[89] Sinnokrot, M. O., and Sherrill, C. D. (2006) High-Accuracy Quantum Mechanical Studies of π-π Interactions in Benzene Dimers. *J. Phys. Chem.* A110:10656-10686.

[90] Saebø, S., and Pulay, P. (1985). Local configuration interaction: An efficient approach for larger molecules. *Chem. Phys. Lett.* 113:13-18.

[91] Furche, F. (2001). Molecular tests of the random phase approximation to the exchange-correlation energy functional. *Phys. Rev.* B64:195120 (1-8).

[92] Eshuis, H., Bates, J. E., and Furche, F. (2012). Electron correlation methods based on the random phase approximation. *Theor. Chem. Acc.* 131:1084 (1-18).

[93] Scuseria, G. E., Henderson, T. M., and Sorensen, D. C. (2008). The ground state correlation energy of the random phase approximation from a ring coupled cluster doubles approach. *J. Chem. Phys.* 129:231101 (1-5).

[94] Parkhill, J. A., and Head-Gordon, M. (2010). A tractable and accurate electronic structure method for static correlations: The perfect hextuples model. *J. Chem. Phys.* 133:024103 (1-11).

[95] Tsuchimochi, T., and Scuseria, G. E. (2009). Strong correlations via constrained-pairing mean-field theory. *J. Chem. Phys.* 131:121102 (1-5).

[96] Booth, G. H., Alavi, A., and Thom, A. J. W. (2009). Fermion Monte Carlo without fixed nodes: A game of life, death, and annihilation in Slater determinant space. *J. Chem. Phys.* 131:054106 (1-11).

[97] Chan, G. K-L., and Head-Gordon, M. (2002). Highly correlated calculations with a polynomial cost algorithm: A study of the density matrix renormalization group. *J. Chem. Phys.* 116:4462-4476.

[98] Shepard, R. (2005). A General Nonlinear Expansion Form for Electronic Wave Functions. *J. Phys. Chem.* A109:11629-11641.

[99] Schaefer, H. F. III. (1977). *Applications of Electronic Structure Theory.* NY: Plenum Press.

In: An Introduction to Electronic Structure ISBN: 978-1-53618-411-2
Editor: Nadia T. Paulsen © 2020 Nova Science Publishers, Inc.

Chapter 5

PERTURBATIVE ACCOUNT OF ELECTRON CORRELATION EFFECTS IN THE INTERNAL ROTATIONAL BARRIER OF MOLECULES: A STATE SPECIFIC STRATEGY

Sudip Chattopadhyay[*]

Department of Chemistry, Indian Institute of Engineering Science and Technology, Shibpur, Howrah, India

Abstract

Based on a concept of different prescription for different correlation, a multireference BrillouinWigner perturbation scheme with improved virtual orbitals (IVO-BWMRPT) has been presented as an accurate and affordable computational protocol for treating the electronic states plagued by quasidegeneracy. It deals with only a single-root problem in the Hilbert space treating all components of the model space on the same footing. The IVO-BWMRPT approach has several attractive properties (explicit size-extensivity and intruder-free nature without using any numerical threshold or *ad hoc* parameter) for investigating various chemical processes and problems. The IVOs were generated using the complete active space configuration interaction (CASCI) scheme and exploited as a means of recognizing state-specific nondynamical (neardegeneracy) effects. In IVO-BWMRPT, given an IVO-CASCI wave function, the re-

---

[*]Corresponding Author's E-mail: sudip_chattopadhyay@rediffmail.com.

maining dynamical correlation can be efficiently recovered using BWM-RPT scheme. Main motivation in using the IVO-CASCI is that it does not need iterations, nor does it face convergence difficulties that may occur in complete active space self-consistent field (CASSCF) estimations, although IVO-CASCI function cherishes the appealing trait of the CASSCF wave function. Investigations of the torsion of diimide and hydrazine demonstrate that biradicaloid electronic structure can be described nicely with this method where the usual single-reference description fails. As the present approach does not utilize any parameter or numerically unstable operation, the surfaces provided by IVO-BWMRPT are smooth all along the reaction path. A promising accordance between the present estimates and literature values has been found. The IVO-based perturbative technique can open the possibility for an accurate description of the energy surfaces for the ground and excited states of small to large systems at the *ab initio* level in a simplified fashion.

1. INTRODUCTION

Due to renewed controversies over the barrier's origin, a clear understanding about the energetics and dynamics of internal rotational or torsional barriers even for small systems is very essential to fathom conformational changes which are fundamentally important in various phenomena of chemical and biological processes such as protein folding and misfolding, signal transduction cascades in cells, as well as chemical reactivity for individual molecular systems. The existence of hindered internal rotation about single bonds, which also is one of the most rudimentary conceptions in conformational analysis of various organic molecules, has been explored both experimentally and theoretically [1]. Electronic quasidegeneracies often encountered in internal rotational processes (specially at twisted region) lead to multiconfigurational (MC) wave functions [which are not dominated by a single configuration, but rather include several leading configurations] that present a challenge to the practitioners of quantum chemistry. Such systems are also called multireference (MR) systems or strongly correlated systems because their electronic structure is best treated with methods called MR strategies. The presence of nondynamical correlation on top of the dynamical one is usually unavoidable when scanning the energy surface over the wide range of geometrical parameters. Most molecules of practical interest attribute electronic structures governed by more than one configuration. Inherently, MR systems require careful account and balancing of both dynamic and static correlation effects and their interference which makes the theoretical investigation of their nature and reactivity difficult. The key methodological problem is the treatment of both correlations that have rather different requirements on the wave function ansatz. For computing static correlation, a wave function ansatz requires high flexibility in configuration space whereas accounting for dynamical correlation, parametrization of the wave function capable of efficiently treating a large set of (virtual) orbitals is very crucial in order to treat excitations out of a qualitatively correct, strongly correlated reference function. The description of molecules in which inter-play of both correlation are crucial stands as a subject in theoretical quantum chemistry that continues to see active persuasion.

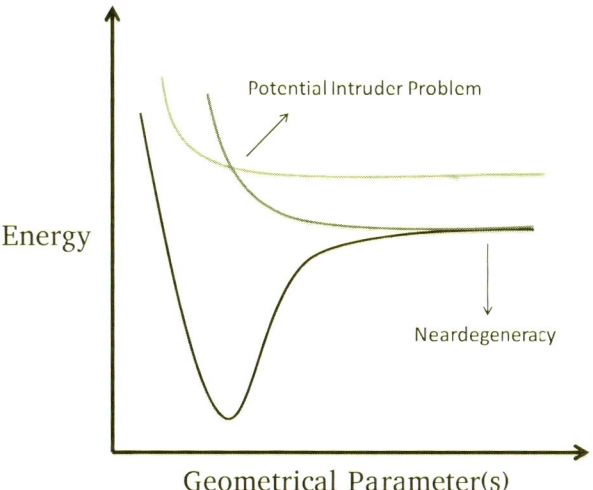

Figure 1. Pictorial representation of appearance of neardegeneracy and intruder-state problem.

As one of the useful methods for treating strongly correlated molecular problems, MR perturbation theory (MRPT) forms an important theoretical framework for a realistic yet practical treatment at the *ab initio* level [2, 3]. Various MRPTs have been proposed over the last few decades many of which have been conveniently reviewed [2, 3] and compared [4, 5] by various workers. Generally, the basic aspect of MRPT consists of two steps: the first step targets primarily the treatment of quasidegeneracies (usually termed nondynamic, static, or strong electron correlation) by considering a set of configurations as references, and the second step involves handling of the remaining electron correlation (dynamic electron correlation) as is done in the single-reference (SR) case. A large number of efficient and effective MRPTs have been developed over the years [6, 7, 8, 9, 10, 11, 12, 13, 14, 15, 16, 17, 18, 19, 20, 21, 22, 23, 24, 25, 26, 27]. Similar to other MR-based methods, all MRPTs require a manual selection of the reference (model) space for the problem of interest. The main dissimilarity between these MRPTs is in the definition of the zeroth-order wave function and Hamiltonian [2, 3]. Several production-level implementations of 'diagonalize-then-perturb' MRPT methods are available in different well-used quantum chemistry packages. A troubling issue of different MRPT methods is the so-called intruder-state problem [emerges when low-lying excited configuration(s) outside of the reference space become quasidegenerate with the high-lying function(s) of the model space, see Figure 1] that invites divergences in the calculations of first-order amplitudes, thereby hampering the utility of the approach seriously. In such a situation, conventional form of the Bloch equation can hardly be converged smoothly, leaving the whole electron correlation problem unsolved. Note that the intruder-state issue is not limited to MRPTs [2, 3], but also appears frequently in MR-based coupled cluster (CC) calculations. In the case of nonperturbative MR-correlated methods, intruder-state problem results from the existence of close-lying multiple solutions to a set of nonlinear equations [iterative protocols yield other solutions of the equations than the desired one] and hence it is difficult to obtain the desired physical solution of the equations due to serious convergence problems. Thus, intruder state problems are a notable hurdle on the way toward routine uses of the approaches. Here, it is worth mentioning that realizing a natural remedy to the convergence problem due to the intruders in MRPT method can provide an understanding about how to design an acceptably faithful MR-correlated scheme that is numerically robust. The intruder state problem can be cured, if not completely removed, in a pragmatic manner through the introduction of some parameters [28, 29]. The failure of MRPT to describe near-degenerate electronic states due to intruder states has motivated the development of various useful practical MRPT schemes which would circumvent the convergence difficulties in a natural manner. An effective way to rectify the intruder problem leads to the emergence of the so-called *state-specific* MRPT (SSMRPT) method orginally introduced by Mukherjee and coworkers [15], where only one state is picked up from the reference space [15, 16, 17, 18, 19, 20, 21, 22]. There are many variants of

SSMRPT method depending on the choices of the reference state, zeroth-order Hamiltonian and perturbative expansion.

We have recently designed an effective Hamiltonian based second-order BrillouinWigner perturbation theory (BWMRPT) with an improved virtual orbitals (IVO) complete active space configuration interaction (CASCI) (IVO-CASCI) reference function for treating strongly correlated systems that are not well characterized by a single-reference description. The resulting approach, termed as IVO-BWMRPT, falls within the Hilbert-space SS category. The IVO-BWMRT equations do not experience numerical irregularities due to intruder states which come across when electronic states that recline within and outside the space comprised by the effective Hamiltonian become quasidegenerate. Similar to other SSMR methods, IVO-BWMRPT is mainly tailored for the ground state. However, excited states can also be obtained as lowest states of an irreducible representation of the molecular point group or spin. From a formal point of view, separate computations converging to different target roots of the effective Hamiltonian are needed in the Jeziorski-Monkhorst SS protocol, while the state universal (SU) calculations (multistate approach) provide a simultaneous estimation of as many electronic roots as the dimension of the reference space (through the full diagonalization of the effective Hamiltonian). IVO-BWMRPT is strictly parameter-free, intruder-state free, and size-extensive method. Size-consistent property with localized orbitals ia another distinctive aspect of this formalism. The complete active space self-consistent field (CASSCF) protocol is commonly applied as the starting point for most MRPT methods, however, in many situations, the calculations at CASSCF level can lead to multiple solutions that invite convergence problems, root-flipping between close-lying states, and symmetry breaking. As an efficient alternative to CASSCF, one can consider the IVO-CASCI scheme [30, 31], in which the IVOs are used in a CAS space. IVO-CASCI scheme avoids the orbital optimization step altogether (that ameliorates some of the problems of the CASSCF approach) without damaging the attractive features of the CASSCF scheme. The key result of IVO-CASCI method is that the it agrees quite well with CASSCF, at a greatly reduced cost, and this extends the usefulness of the IVO-CASCI scheme to larger systems. The IVO-CASCI does not entail iterations, nor does it experience numerical divergence which may appear in CASSCF computations. However, well-balanced CAS-based correlated computations are in practice limited to the investigations of small-to-medium molecules as the size of the CAS expansion grows exponentially as the active space is increased. Recently, various approaches have been proposed for treating strongly correlated systems at much lower computational costs compared to CAS-based MR calculations [23, 24, 32, 33].

Continuing the recent endeavor of our group, here, internal rotational barriers of the NN bond in 1,2-diazene or diimide (N_2H_2) and hydrazine (N_2H_4) will be investigated. Neardegeneracy effects are usually supplicated to describe the complex essence of the systems. Owing to their small size, they are often used as probing grounds for an unambiguous understanding of conformational behaviors of molecules that contain the N-N bond(s). As a result, various investigations on these systems have been performed both experimentally and theoretically using different *ab initio* methodologies. Various researchers argued that the internal rotation barriers of such molecules originated due to an intricate inter-play of both the hyperconjugation and steric repulsion effects. Nevertheless, these two systems have been used frequently to judge MR-electronic structure methods tailored to handle near-degeneracy. The nondynamical electron correlation effects are significant near the twisted form with the dihedral-angle parameter of around 90°.

2. THEORY: A BRIEF RESUMÉ

A detailed analysis along with emergence of the SSMRPT method with Brillouin-Wigner expansion can be found in the previous papers [15, 19, 34]. Without the details of the derivation and manipulations, the basic underlying theoretical issues of the multireference, state specific, second-order, Brillouin-Wigner perturbation theory with IVOs (IVO-BWMRPT) is presented here in order to highlight the necessary notation and terminology of the procedure [34]. The BWMRPT strategy is employed to recover the missing dynamical correlation neglected at the level of IVO-CASCI calculations. The present method belongs to the category of 'diagonalize-then-perturb-then-diagonalize' schemes whereas widely used MRPT methods such as MRMP2, CASPT2 and NEVPT2 falls within the category of 'diagonalize-then-perturb' protocols.

Before describing the methodological aspect of IVO-BWMRPT, it is useful to describe the underlying philosophy behind the generation of IVO-CASCI orbitals which are subsequently used in BWMRPT procedure. The IVO-CASCI procedure can be employed as an approach of its own for computing different electronic states, but also as an avenue that yields the orbitals required in the correlated MR-computations. An IVO-CASCI wave function preserves the main merit of the CASSCF. It is well defined, and has an upper bounds nature to the energies of the states. The conventional CI scheme yields both the occupied and unoccupied orbitals as well as their energies using a *single* V^N potential Fock operator (where N is the number of electrons present in the reference HF function). On the other hand, the IVO-CASCI uses *multiple Fock* operators to describe the valence orbitals and the corresponding energies from a $V^{()N-1)}$ potential Fock operator in order to optimize the CASCI and hence to minimize the higher order perturbative corrections. Note that, the IVOs treat all valence orbitals democratically. In HF calculations, on the other hand, $V^{()N-1)}$ potential is used for the description of the orbitals occupied in the reference HF state and V^N potential for the valence orbitals that are unoccupied in the reference state. It should be emphasized that the CASSCF calculations effectively associate a CASCI step using orbitals optimized for a target state or for some weighted average of several states. The IVO-CASCI is a full CI in a given reference (active) space and hence, the partial orbital optimization effects can be encompassed in IVO-CASCI if the CAS is appropriately selected. The lower computational cost of the IVO-CASCI over the CASSCF method is evident from the properties of the former: IVO-CASCI calculations start with a single ground state HF procedure, followed by the diagonalization of a single Fock operator, and eventually a CASCI step. The IVO-CASCI wave functions are then at hand for subsequent refinement of the wave function employing MRPT computations, here state-specific BWMRPT. Note that the IVO protocol encompasses just the first two steps. No iterations are needed, and various CASs may be handled without reiterating the above mentioned first and second steps. Note that IVO-CASCI neglects most dynamical electron correlation effects, and it occasionally yields qualitatively different findings compared to higher levels of dynamically correlated approaches. Thus, like the CASSCF, IVO-CASCI computation would usually be followed by a perturbative calculation of the missing dynamical correlation effects.

The BWMRPT strategy is employed to recover the missing dynamical correlation neglected at the level of IVO-CASCI calculations. The naive description of dynamical correlation effects in MR situations is extremely challenging. In IVO-BWMRPT method, the Schrödinger equation is converted into an effective eigenvalue problem as:

$$\sum_\nu \widetilde{H}^{(2)}_{\mu\nu} c_\nu = E^{(2)} c_\mu \qquad (1)$$

Here, $\langle\phi_\mu|H|\phi_\nu\rangle = H_{\mu\nu}$, and $\langle\chi_l^\mu|H|\phi_\mu\rangle = H_{\mu l}$. BWMRPT works with a CAS, and treats each of the model space functions on the same footing by exploiting state-specific parametrization of the state-universal Jeziorski-Monkhorst (JM) ansatz: $|\Psi\rangle = \sum_\mu \exp(T_\mu)|\phi_\mu\rangle c_\mu]$ [35]. Note that the set of configurations, $\{\phi_\mu\}$constitute the reference space, CAS. It is stressed here that the selection of the functions, ϕ_μ included in CAS is dictated by the nature of the problem under study. In JM-ansatz, the cluster operators T_μ are described as regular excitation operators relative to the reference configuration ϕ_μ and is specific for each reference. The expansion coefficients c_μ are a *priori* unknown. Eq. (1) indicates that the CI coefficients of wave function as well as second-order energy for the target state can be evaluated by the diagonalization of the second-order effective Hamilton $\widetilde{H}_{\mu\nu}^{(2)}[= H_{\mu\nu} + \sum_l H_{\mu l} t_\nu^{l(1)}(\nu)]$. But to construct the effective hamiltonian matrix, knowledge of the cluster amplitudes is also required. The first-order perturbation equations for the amplitudes of the target state can be obtained by employing suitable sufficiency condition to the electronic Schrödinger equation (introduced with the aim of matching the number of equations and unknown cluster amplitudes) that can be expressed as

$$t_\mu^{l(1)}(\mu) = \frac{H_{l\mu} + \sum_\nu^{\nu\neq\mu} t_\mu^{l(1)}(\nu)\widetilde{H}_{\mu\nu}^{(2)}(c_\nu/c_\mu)}{(E^{(2)} - \widetilde{H}_{\mu\nu}^{(2)}) + (E_{0,\mu\mu} - E_{0,ll})} \qquad (2)$$

Here, $\chi_l^\mu|T_\nu^{l(1)}|\phi_\mu\rangle = t_\mu^{l(1)}(\nu)$ and $t_\mu^{l(1)}(\mu)[= \langle\chi_l^\mu|T_\mu^{l(1)}|\phi_\mu\rangle]$. Note that appropriate sufficiency conditions are necessary to be invoked to reach the above equation [15], which make the number of equations equal to the number of variables required to solve. The cluster amplitude finding equation contains two types of terms termed as the direct and the couplings. Here, excitation operators were limited to single and double. χ_l describe the excited configurations with respect to reference up to excitation rank considered here, excluding other reference functions obtained by internal excitations. Note that when the JM ansatz is invoked in the Schrödinger equation and projected on the excited functions originated by the cluster operator for each reference function, the number of equations that emerge is less than the number of unknowns in the JM ansatz. To overcome this problem, *suitable supplementary* or *supplementary conditions* need to be imposed for solving the amplitude equations. It is worth stressing that the amplitude finding equation of IVO-BWMRPT explicitly carries the CI coefficients displaying the state-specific nature of the method. This reference coefficient weighing emerges from the sufficiency condition used to derived the equations. The choice of supplementary conditions is not unique and accordingly, various theories of SSMR formulation have emerged [18]. Hubač and co-workers [18] also suggested another single root BW-based MRPT imposing suitable sufficiency condition in Bloch generalized equation. The amplitude equations in BWPT of Hubač and co-workers are naturally decouple for each reference function and do not contain reference space coefficients. The amplitude finding equations of this method are coupled only through the exact energy of the target state and thus the transition density matrix elements connecting the model space function and virtual space function are not required at all that make the BWPT method computationally more cost effective compared to the present IVO-BWMRPT. However, the price paid is the lack of size extensivity of the method which is the strong objection to the choice of sufficiency conditions used by Hubač and co-workers [18].

The stability of the denominators in presence of intruders in Eq. (2) is quite evident from the fact that the desired energy E always remains well-separated from the energies of the virtual functions. From the viewpoint of perturbation scheme, intruders correspond to excited configurations with vanishing energy denominators. Therefore, IVO-BWMRPT circumvents intruder state problem without using arbitrary level-shift technique that regularizes the offending denominators. It should be emphasized that the widely used MRPT methods such as MRMP2 and CASPT2, as well as their multiroot versions suffer from the intruder state problem which prohibits stable energy convergence of the working equations. This is usually likely to occur when the CAS/CMS is used. Containing only connected terms in Eqs. (2) and (1) with a complete model space (CMS) guarantees an explicit size-extensivity of the method. Therefore, the wave function based IVO-BWMRPT method exploits the philosophy of the effective Hamiltonian formalism which uses a reference space, given by a linear combination of configuration state functions describing a good zero-order approximation for the desired target states. The first-order amplitude equations Eq. (2) along with the equation for the target second-order energy [Eq. (1)] characterize the IVO-BWMRPT method. However, it should be noted that the perturbative energy for the BWMRPT scheme is *pseudo*-second-order in nature as the amplitude finding equations are not truly first-order due to the presence of explicit $\widetilde{H}_{\mu\nu}^{(2)}$ in an amplitude equation, Eq. (2). As the cluster amplitudes finding equation depends on the desired energy, Eqs. (2) and (1) need to be solved self-consistently in the present formulation. Note that if the energy emerging from Eq. (1) through diagonalization of $\sum_\nu \widetilde{H}_{\mu\nu}^{(2)}$ differs from the previous iteration energy by less than the given threshold, we stop the procedure. This iteration process is commenced by diagonalizing the Hamiltonian matrix defined in the reference space:

$$\sum_\nu H_{\mu\nu} c_\nu^0 = E_{\text{CAS}} c_\mu^0 \qquad (3)$$

A defect common to the commonly used MRPT methods such as MRMP2, CASPT2, NEVPT2, and so on is that the coefficients of the configurations which build up the zero order variational wave function are left unchanged upon use of the first order perturbation. Such contracted description can be conspicuous in avoided crossings or in excited/ionized states of mixed valence-Rydberg nature which involve strong mixing between different zero-order configurations. The desired modification of the reference space coefficients can take place in the present IVO-BWMRPT method with the use of diagonalization processes via Eq. (1) that leads to full relaxation of the nondynamical correlation effect where the reference coefficients of the active functions are iteratively updated as dynamical correlation is encompassed. This brings in the reference-specific orbital relaxation in the calculation. This also makes the approach rather less sensitive to nature of the orbital and the active space employed. The unrelaxed description of the IVO-BWMRPT follows naturally if one uses the frozen coefficients for the reference functions akin to CASPT2, NEVPT2 and MRMP2. Theerefore, the reference space components can be either iteratively revised or fixed at the IVO-CASCI values in IVO-BWMRPT method as that of its parent-version. In that sense, it is a flexible MR-method. Here, the results of relaxed version of IVO-BWMRPT have been presented. The difference between various MRPT methods lies in the contraction of the basis states that span the first-order wave function corresponding to the target state(s). Here it is worth noting that the basis present in strongly contracted NEVPT2 formalism has a higher level of contraction than the CASPT2 or MRMP2 ones. Solution of Eq. (1) provides only one electronic state at a time which is supposed to be exact in the limit (the electronic state of interest). All other roots are generated as byproducts of the process, serving to maintain a sufficient gap (buffering) of the root of interest and getting rid of the intruder states by bribing them away. Generally, such a single-root separation removes numerical divergence of first-order amplitudes due to the intruder state problem. It is worth stressing that an N-dimensional effective operator is build in the effective Hamiltonian schemes and N-target states corresponding to the original Hamiltonian H is generated via diagonalization process. In the case of spin-free IVO-BWMRPT formulation, where the cluster operators are defined in terms of the spin-free unitary generators, the Fock-like operator is defined with respect to $\phi_{0\mu}$ which is the largest closed-shell component of the corresponding reference configuration, ϕ_μ. Then IVO-BWMRPT method is applicable for the reference space of arbitrary spin multiplicity. The detail discussion of the spin-adaptation scheme for BW-based SSMRPT protocol is not presented here and one can refer to the review article [2, 36] for the purpose. The aforementioned desirable properties of IVO-BWMRPT method make it a serious applicant for coherent administering of strong electron correlation effects in MR-approaches and hence it has emerged as a suitable MR quantum chemistry method for computing energy surfaces of excited electronic states as well as ground states. Structural nature of the IVO-BWMRPT guarantees accurate descriptions of variation of neardegeneracies of the electronic states involving bond-dissociation processes [34]. By using proper root-homing procedure, one can access ground or excited electronic state in IVO-BWMRPT calculations.

The IVO-BWMRPT scheme address some of the important objections inherent to CAS-based MR methods, including the unfavorable scaling with respect to the number of active orbitals, and numerical instability stemming from an explicit division by the model space coefficients in the cluster finding equation. Though such convergence issues can be reduced by regularization schemes or by simply dropping the cluster amplitudes corresponding to problematic reference configurations, however, no general prescription has been reported in the literature. Each system generally needs a special consideration, thereby hindering the applicability of the protocol as a black-box MR-approach. Moreover, these remedies may also yield discontinuities or bumps in the computed surfaces. This difficulty is common to the case of any method based on JM-ansatz of the wave function whenever an implied division by reference coefficient appears in the working equations. Although treating all reference functions on the same footing, all HS-MR schemes including BWMRPT are not invariant with respect to orbital rotations. It should be noted that as more and more dynamical electron correlation effect is incorporated in the effective Hamiltonian mentioned above, MR-schemes become less sensitive to orbital relaxation effects. Therefore, one can keep employing the HF-like orbitals and skip the expensive orbital optimization step. Additionally, there has been no analytical gradient protocol for SSMRPT with both RS and BW expansion schemes, and consequently, the applicability of the method towards the exploration of energy surface using geometry optimization scheme is somewhat limited. For these reasons, the conception of MR methods (satisfying the properties of a good many-body method, while remaining practicable and sufficiently accurate) remains a challenging area of research in the realm of electronic structure theory.

3. RESULTS AND DISCUSSION

The application of the IVO-BWMRPT procedure described above has been illustrated by considering the internal rotational barrier about the NN bond for molecules such as diimide and hydrazine. Basis sets employed here are taken from the EMSL database [37]. Eqs. (1) and (2) describe the working equations for second-order IVO-BWMRPT method, a robust and efficient scheme for the electron correlation problem for systems requiring a MR strategy, by means of which one can study correctly the internal rotational surfaces considered here. This method can be employed to large molecules using appropriate large basis sets (to gain desired accuracy). Moreover, the refinement of the method can be done either by considering the perturbation series to higher order, or by refining the reference function [20, 21, 34]. Our in-house IVO-BWMRPT code is interfaced with the GAMESS (General Atomic and Molecular Electronic Structure System) programs and exploits the IVO-CASCI procedure of the package to construct reference function.

3.1. 1,2-Diazene or Diimide: N_2H_2

This system is isoelectronic with the ethylene molecule which has already been studied by us using the IVO-SSMRPT [Rayleigh-Schrödinger (RS) version of SSMRPT method IVO-CASCI reference function] [34]. The method has been widely applied to study radicals, chemical reactions, and electronic spectra [20, 21, 22]. The *cis/trans* isomerization of N_2H_2 can occur by either an inversional or a rotational mechanism. Recently, Sand *et al.* [38] stated that MR-wavefunction schemes suggest the rotational pathway to be energetically more favorable while SR-approaches recommend the inversional pathway to be more favorable. An illustrative application of IVO-BWMRPT to the*cis/trans* isomerization reaction of N_2H_2 through rigid rotation about the NN double bond is addressed here. Diimide is often employed as a reagent in stereospecific hydrogenation of double bond. The mechanism of hydrogenation needs isomerization from the stable *trans*-form to *cis* one, which is believed to be the rate

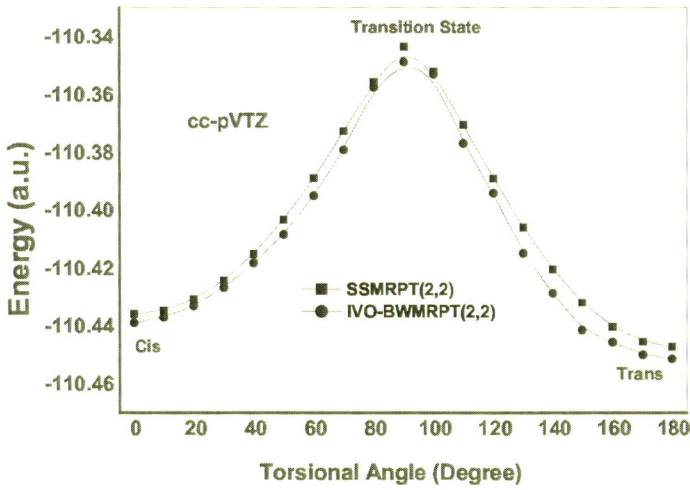

Figure 2. Twisting energy surface of N_2H_2 (due to the rigid rotation about the central N-N bond) as obtained by IVO-BWMRPT(2,2)/cc-pVTZ method.

determining step of the hydrogenation process. Therefore, information about rotational energy surface for diazene is a crucial step towards the analysis of elementary reactions associated with such processes. There have been various theoretical [38, 39, 40, 41, 42, 43, 44, 45] and experimental [46] investigations on the diazene system. In fact, the ground state of the diazene is primarily dominated by a single configuration [$\phi_1 = 1a^2 1b^2 2a^2 2b^2 3a^2 3b^2 4a^2 5a^2$ for *trans* and $\phi_2 = 1a^2 1b^2 2a^2 2b^2 3a^2 4a^2 3b^2 4b^2$ for *cis*] in the regions of local minima. Note that they differ by double excitation. At the twisted geometry, both ϕ_1 and ϕ_2 are important: a typical MR situation. In true sense, the rigid rotation of the diazene describes a typical two-state problem in which the weights of ϕ_1/ϕ_2 can vary from 0 to 1 in a continuous fashion. Therefore, the rigid rotation in the 1,2-diimine molecule yields to a crossing of occupied and unoccupied orbitals of different symmetries, which, according to the Woodward-Hoffman rules, are not allowed. Here, the internal coordinates such as NH= 1.0320 Åand NN= 1.2501 Åbond lengths and $\angle NNH$ =111.71° were held fixed at their experimental values for the *trans*-conformation [46]. IVO-BWMRPT calculations with cc-pVTZ basis set have been carried out as a function of the dihedral angle, $\theta = \angle H - N - N - H$. In the present study, we employed CAS(2,2) consisting of two closed shell-type configurations, ϕ_1 and ϕ_2. The $1s$ core orbital corresponding to the N atom has been kept frozen in the correlation treatment. Ground-state energies obtained using IVO-BWMRPT(2,2)/cc-pVTZ at various torsional angles are described in Figure 2. It can be seen from the figure that the IVO-BWMRPT approach furnishes a smooth and correct shape of the energy surface along the entire spectrum of torsional coordinates including nondegenerate zones without producing any unphysical cusp even at dihedral angle near 90°. This fact indicates that IVO-BWMRPT method is very useful to scan stationary points on torsional surface without altering the CAS. This aspect is beneficial to estimate relevant energetics faithfully when chemistry in polyatomic molecules is assessed. The present IVO-BWMRPT method performs as well as the high-level MR methods such as sr-MRCCSD/cc-PVTZ [44],]sr-BWCSSD/[4s3p] [40], and MRCI/aug-cc-pVQZ [41] indicating the strength of the IVO-BWMRPT method. The IVO-BWMRPT method, like the other high-level correlated-MR protocols, is able to switch itself from a nondegenerate to a quasidegenerate situation in a continuous fashion and thus prevents an abrupt change in the torsional energy surface. Note that although Hilbert space MR-CCSD (multireference coupled cluster with singles-doubles) calculations provide the correct topology of the energy barrier almost over the entire range of the torsional angles, they do not furnish a good description in the highly nondegenerate regions [in the close proximity of 0° (*cis*-conformer) and 180° (*trans*-conformer)] due to the intruder state problem. As SU-MRCC (state-universal MRCC)and its perturbative counterpart are multi-root techniques, any inappropriate treatment of one state due to potential intruders may ruin the convergence behavior of the equations for other states. Although, DIP-EOMCC (double ionization potential equation of motion CC) method provides smooth and cusp-free torsional surface, the standard CC protocols fail to handle the energy surface in a correct manner in the whole range of dihedral angle [42]. It is worth stressing that CR-CC(2,3) method also failed to give a smooth energy profile for cis-trans isomerization process of N_2H_2 [43].

To judge the quality of the IVO-BWMRPT energy surface, the barrier height [can be identified as relative energies of *trans* with respect to the *cis* and *transition* state] has also been estimated. It should be emphasized that a proper representation of transition states (which is often associated with significant effects of the nondynamical electron correlation) of chemical reactions is the most important aspect for quantitative understanding of quantum chemical processes. Table (1) assembles the energetic barriers for the *trans*→*cis* isomer-

Table 1. Barrier to internal rotation (kcal/mol) and relative energies of *cis*-form with respect to *trans*-one for the isomerization process of 1,2-diazene

Ref.	Method	TS-N_2H_2	*cis*-N_2H_2
Present	**IVO-BWMRPT(2,2)/cc-pVTZ**	64.95	7.39
Ref. [45]	CASSCF-SSMRPT(2,2)/cc-pVTZ	64.24	6.98
Ref. [43]	IVO-MRMP2(2,2)/aug-cc-pVTZ	48.45	5.87
Ref. [41]	MRCI+Q(12,12)/aug-cc-pVQZ	54.96	5.05

ization, and the barrier for the transition state (TS-N_2H_2) corresponding to the rotational pathway. The accuracy of the IVO-BWMRPT rotational barriers may further be calibrated by comparing with the estimates of previous works. It should be noted that the barriers to rotation from the CC approaches are in the range 65.8-71.9 kcal/mol as reported by Sand-Schwerdtfeger-Mazziotti [38]. For *trans-to-TS-N_2H_2* rotational pathway, the IVO-BWMRPT barrier of 64.95 kcal/mol in the cc-pVTZ basis set is similar to that previously calculated by SS-MRPT(2,2)/cc-PVTZ scheme [45]. The reference states of SSMRPT [45] have been obtained from the state specific CASSCF computations. Thus, one can note that the use of the IVO-CASCI orbitals in lieu of the CASSCF ones lead to almost similar estimates. Note that the CCSD, CRCC(T), and CCSD(T) level of calculations yield rotational barriers of 67.8, 62.6, and 60.0 kcal/mol, respectively [38] which are in close agreement with the present estimate. The barrier to the internal rotations of 65 kcal/mol provided by DIP-EOM-CCSD/cc-pVTZ level [42] is also in nice agreement with the IVO-BWMRPT(2,2) value. The corresponding barrier value for the CCSD/cc-pVTZ is 91.8 kcal/mol [42] which is significantly higher than the values provided by other methods discussed here indicating the inability of the CCSD method to handle the twisted region of *trans*→*cis* isomerization pathway. The single-reference CC theory yields a cusp near twisted region due to the change in the Hartree-Fock reference function. In contrast, varying the torsional angle near transition state in the DIP-EOM-CC method does not demonstrate the unwanted cusp. Thus, the DIP-EOM-CC method provides a useful way to account for rotational energy surfaces in the situations where the reference functions is not any longer described by a single configuration function. The barriers 52.7 and 52.6 kcal/mol, predicted by MRPT2 and ACSE (anti-Hermitian contracted Schrödinger equation) methods, respectively [38] are higher than the IVO-BWMRPT value by around 12 kcal/mol. The reaction barriers from the parametric variational 2-RDM/aug-cc-pVDZ [38] and MRCI+Q(12,12)/aug-cc-pVQZ schemes [41], respectively are at least 14 kcal/mol smaller than those from the IVO-BWMRPT method. The rotational barrier heights from these methods such as 2-RDM and MRCI+Q(12,12) are 51.3 and 54.96 kcal/mol, respectively. IVO-MRMP2 predicts a much smaller rotational barrier of 48.45 kcal/mol compared to the IVO-BWMRPT and CC-methods discussed here.

The energy gap between *cis* and *trans* comes out close to 7.39 kcal/mol at the IVO-BWMRPT/cc-pVTZ compared to 5.87 and 5.05 kcal/mol for the IVO-MRMP(2,2)/aug-cc-pVTZ [43] and MRCI+Q(12,12)/aug-cc-pVQZ [41], respectively. The IVO-BWMRPT result stay close with the value of 7.7 kcal/mol obtained at the DIP-EOM-CCSD/cc-pVTZ level of calculations [42]. The CCSD value of 5.3 kcal/mol [42] is in good agreement with the IVO-MRMP2 and MRCI+Q calculation. This is due to the fact that *trans* and *cis* conformers of 1,2-diazene are well described by the traditional SRCC method.

Note that IVO-BWMRPT supplies electronic energies that are as good as those obtained from other high-level MR-calculations. IVO-BWMRPT produces a more accurate barrier than that from CCSD scheme at a lower computational cost. The present investigation argues that an approach describing the *cis/trans* isomerization must have an ability for balanced treatment of both SR and MR situations. As that of other highly established many-body methods, IVO-BWMRPT energy gap for *trans→cis* isomerization predicts that the *trans*-form is energetically more stable than the *cis*-one. As illustrated in Table (1), the *cis-trans* energy difference obtained by the IVO-BWRPT is acceptably close in proximity to those of SS-MRPT(2,2), IVO-MRMP2(2,2), and MRCI+Q(12,12) estimates. Here, it should be mentioned that the CAS employed in MRCI by Biczysko *et al.*[41] is the full valence CAS [consisting of 12 active electrons in 10 active orbitals] in comparison to our CAS(2,2).

3.2. Hydrazine: N_2H_4

The next system investigated in this article is the rotational barriers of the NN bond in hydrazine rather than the barriers of inversion of a NH_2 group. As that of N_2H_2, hydrazine is a chiral molecule. To understand conformational natures of chemical and biochemical compounds consists of the N-N bond,

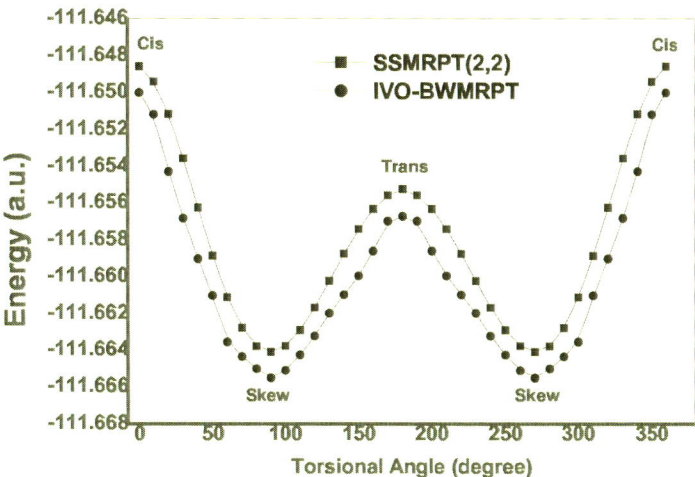

Figure 3. Energy surface of the ground state hydrazine as a function of torsional angle as obtained by IVO-BWMRPT/cc-pVTZ scheme.

N_2H_4 has been substantially investigated as a model molecule both by theory [47, 48, 49, 50, 51] and experiment [52, 53, 54, 55, 56]. Figure 3 shows the rotational energy profile of hydrazine as a function of dihedral angle as obtained by IVO-BWMRPT(2,2)/cc-pVTZ. Optimized structures used here have been determined by numerical gradient IVO-SSMRPT(2,2)/cc-pVDZ scheme [57]. As expected, the IVO-BWMRPT provides smooth, cusp-free behavior around the twisted region where the various methods often invite discontinuity. Figure 3 indicates that the hydrazine passes through two transition states [at the dihedral angles of 0° and 180°] in the course of internal rotation around the N-N bond of the system, which is consistent with the observation of other workers. Thus, the internal rotational pathway in N_2H_4 is experienced by the two barriers: two energetically favored rotational positions. As a result of this, torsional surface of hydrazine exhibits a double well. The computed double well energy surface recommends that the *cis*-conformer is energetically much higher than the *trans*-one as that of N_2H_2, which emerges from the nuclear-nuclear Coulombic repulsion between outer H/H atoms in *cis*-form. In *cis*-form, the hydrogen atoms and the electron lone pairs eclipse each other and hence *cis*-geometry is somewhat less stable than *trans*. The literature argues that the mutual inter-play of both hyperconjugative and steric effects lower the energy of *skew* form and finally form low barrier from *skew* ↔ *trans* and high barrier from *skew* ↔*cis* in hydrazine [47]. In hydrazine, *skew*-structure is primarily the energy-minimum conformer and even energetically more stable than its *trans* form which may have the minimal steric repulsion between the two NH_2 groups from the view-point of simple repulsion model. Here, it is worth stressing that the *skew*-geometry of hydrazine possesses higher hyperconjugative stability effect compared to its *trans*-form, strongly suggesting the preference of the *skew* conformer over both the *cis*- and the *trans*-forms. The values of barriers are collected in Table (2) for various level of calculations. Experimental values and some other theoretical estimates available in the literature are also tabulated for comparison. Table (2) demonstrates that there is a good agreement of IVO-BWMRPT barrier with the literature values [50, 48, 49, 57] indicating the usefulness of the present method. Note that compared with diimide, hydrazine has a much smaller rotational barrier. The IVO-BWMRPT(2,2)/cc-pVTZ yields *trans*↔*skew* and*cis* ↔*skew* barriers of 2.56 and 8.25 kcal/mol, respectively. The corresponding values provided by IVO-SSMRPT(2,2)/cc-pVTZ level are 2.48 and 8.48 kcal/mol, respectively which are in very good agreement with the IVO-BWMRPT values. IVO-BWMRPT(2,2)/cc-pVTZ results are comparable to 2.14 and 8.10 kcal/mol at the CCSD(T)/cc-pVTZ level [48]. The CCSD(T)/cc-pVQZ scheme leads to the barrier values of 2.52 and 8.26 kcal/mol for *trans*↔*skew* and*cis*↔*skew* reaction path. Notice that the energy barriers for *trans* and *cis* forms obtained at the SSMRCCSD(2,2)/cc-pVTZ level are 2.44 and 8.37 kcal/mol in agreement with the IVO-BWMRPT. The far-IR spectrum [52] provides a *trans* barrier of 3.14 kcal/mol for hydrazine in agreement with the present IVO-BWMRPT/cc-pVTZ calculations with a deviation of 0.58 kcal/mol. The corresponding errors are 0.70 and 1.0 kcal/mol yielded by SSMRCCSD and CCSD(T) methods with cc-PVTZ basis set. The *trans*-configuration is favored over the *cis*-form by 5.7 kcal/mol at the level of IVO-BWMRPT(2,2)/cc-pVTZ which is comparable and equally close to the reference SSMRCCSD [57] and CCSD(T) [48] values. Present findings illustrate that recently developed IVO-BWMRPT code is very promising for electronically challenging molecules suffering from neardegeneracy problems, where the conventional *ab initio* schemes cannot provide an accurate description.

Finally, it is worth stressing that in situations/processes carrying a noticeable MR nature, the wave function cannot be acceptably approximated by a single configuration or determinant even from the mathematical point of view.

Table 2. Relative energies or rotational barriers (kcal/mol) of hydrazine as obtained by various methods

Reference	Method	*trans*	*cis*
Present	**IVO-BWMRPT(2,2)/cc-pVTZ**	2.56	8.25
Ref. [57]	IVO-SSMRPT(2,2)/cc-pVTZ	2.48	8.48
	SSMRCCSD(2,2)/cc-pVTZ	2.44	8.37
Ref. [50]	CCSD(T)/aug-cc-pCVTZ	2.13	7.74
Ref. [48]	CCSD(T)/cc-pVTZ	2.14	8.10
	CCSD(T)/aug-cc-pVTZ	2.43	8.19
	CCSD(T)/cc-pVQZ	2.52	8.26
Ref. [49]	B3LYP/aug-cc-pvDZ	2.75	8.54
Ref. [52]	Far-IR Experiment	3.14	-

The MR nature of the reference (zeroth-order) wave function should be considered in such cases. At the same time, almost all MR approaches need a manual selection of the reference space (CAS) for the problem of interest. Another fundamental issue arises from the fact that the contribution of configurations in the reference space can vary along the energy surfaces. The reference configurations contribute significantly at some point on the energy surface and can lose their impact at some other region of the surface indicating that a fixed form of the reference space may invite extra calculations corresponding to these unimportant reference configuration(s) [whose impact on the target wave function is comparable or even smaller compared to some configurations from the (first-order interaction) orthogonal space] when the MR nature of the target wavefunction is lost. We, practitioners of electronic structure theory, hope that a fruitful evasion of all the difficulties associated with genuine MR methods (*vide supra*) will definitely extend our understanding about the nature of various chemically challenging processes at the quantitative level.

CONCLUSION

Rotational barriers, rendered as a crucial issue of conformational analyses in different fields, have various implications. The rotational barrier surfaces of even quite simple systems (such as the internal rotation around the CC bond in ethane) are not entirely apprehended. Our recently suggested IVO-BWMRPT method that combines the advantages of the IVO-CASCI and state-specific BWMRPT (describes dynamical correlation up to the second order) schemes has been presented as a stable and reliable tool to provide a correct description of internal rotational barriers of diimide and hydrazine as the method is able to treat static and dynamic correlation effects in a balanced, accurate, and inexpensive way. The systems considered here carry a reasonable MR nature for the twisted conformations and thus, the standard SR-methods fail to explain correctly the torsional surface. Present method is endowed with most preferable formal characters such as size-extensivity and absence of numerical divergence due to the intruder states. In the present methodology, the reference coefficients can be iteratively updated (relaxed) by solving the eigenvalue problem in the reference space. The energy provided by the method is strictly separable when the orbitals are localized on the disjoint fragments. Another important property of the IVO-BWMRPT approach is its ability of a continuous transition between the single- and multi-reference schemes. In IVO-BWMRPT, the perturbation space can feed back to the active space iteratively within the framework of effective Hamiltonian. It gives stable estimates with respect to variations in the basis set or in the CAS [34]. Present IVO-BWMRPT formulation provides smooth and cusp-free torsional energy surfaces of correct shape for all dihedral angles of the systems treated here. The single-reference formulations usually manifest a cusp near transition state emerging from an abrupt alteration in the Hartree-Fock reference function. IVO-BWMRPT computations on hydrazine yield a smooth double well torsional energy surface where *skew* isomer is energetically favored over *trans* form and supports the *cis* structure as the energy-maximum state which is consistent with previous high-level established calculations. The barrier involved in the diimide isomerization calculated at IVO-BWMRPT level agrees well with the best theoretical values. As a final comment, it should be mentioned that the IVO-BWMRPT methodology offers an attractive avenue with a manageable cost/accuracy ratio to scan energy surfaces in the situations where the starting functions is no longer described by a single configuration.

ACKNOWLEDGMENT

Financial support of the present work came from CSIR [Ref. No. 01(2973)/19/EMR-II] of India.

REFERENCES

[1] Mo, Y., Gao, J., 2007. Theoretical analysis of the rotational barrier of ethane. *Acc. Chem. Res.* 40, 113-119.

[2] Chattopadhyay, S., Chaudhuri, R. K., Mahapatra, U. S., Ghosh, A. Sinha Ray, S., 2016. State-specific multireference perturbation theory: development and present status. *WIREs: Comput. Mol. Sci.*, 6, 266-291.

[3] Lischka, H., Nachtigallová, D., Aquino, A. J. A., Szalay, P. G., Plasser, F., Machado, F. B. C., Barbatti, M., 2018. Multireferenc approaches for excited state of molecules. *Chem. Rev.* 118, 7293-7361.

[4] Chaudhuri, R. K., Freed, K. F., Hose, G., Piecuch, P., Kowalski, K., Wloch, M., Chattopadhyay, S., Mukherjee, D., Rolik, Z., Szabados, Á., Tóth, G., Surján, P. R., 2005. Comparison of low-order multireference many-body perturbation theories. *J. Chem. Phys.* 122, 134105.

[5] Hoffmann, M. R., Datta, D., Das, S., Mukherjee, D., Szabados, Á., Rolik, Z., Surján, P. R., 2009. Comparative study of multireference perturbative theories for ground and excited states. *J. Chem. Phys.* 131, 204104

[6] Hirao, K., 1992. Multireference Møller-Plesset Method. *Chem. Phys. Lett.* 190, 374-380. Nakano, H.; Nakayama, K.; Hirao, K.; Dupuis, M., 1997. Transition State Barrier Height for the Reaction $H_2 CO \rightarrow H_2 + CO$ Studied by Multireference Møller-Plesset Perturbation Theory. *J. Chem. Phys.* 106, 4912-4917.

[7] Andersson, K., Malmqvist, P. A., Roos, B. O., Sadlej, A. J., Wolinski. K., 1990. Second-order Perturbation Theory with a CASSCF Reference Function. *J. Phys. Chem.* 94, 5483-5488.

[8] Angeli, C., Cimiraglia, R., Evangelisti, S., Leininger, T., Malrieu, J.-P., 2001 Introduction of n-electron Valence States for Multireference Perturbation Theory. *J. Chem. Phys.* 114, 10252-10264.

[9] Rolik, Z., Szabados, Á., Surján, P. R., 2003. On the Perturbation of Multi-configuration Wave Functions. *J. Chem. Phys.* 119, 1922-1928.

[10] Hoffmann, M. R., 1996. Canonical Van Vleck quasidegenerate perturbation theory with trigonometric variables. *J. Phys. Chem.* 100, 6125-6130.

[11] Xu, E., Li, S., 2013. Block correlated second order perturbation theory with a generalized valence bond reference function. *J. Chem. Phys.* 139, 174111.

[12] Casanova, D., Head-Gordon, M., 2009. Restricted active space spin-flip configuration interaction approach: theory, implementation and examples. *Phys. Chem. Chem. Phys.* 11, 94779-94790.

[13] Chaudhuri, R. K., Freed, K. F., Chattopadhyay, S., Sinha Mahapatra, U., 2008. Potential energy curve for isomerization of N_2H_2 and C_2H_4 using the improved virtual orbital multireference Møller-Plesset perturbation theory. *J. Chem. Phys.* 128, 144304.

[14] Chattopadhyay, S., Chaudhuri, R. K., Sinha Mahapatra, U., 2008. Application of improved virtual orbital based multireference methods to N_2, LiF, and C_4H_6 systems. *J. Chem. Phys.* 129, 244108.

[15] Sinha Mahapatra, U., Datta, B., Mukherjee, D., 1999. Molecular applications of a size-consistent state-specific multireference perturbation theory with relaxed model-space coefficients. *J. Phys. Chem. A* 103, 1822-1830.

[16] Sinha Mahapatra, U., Chattopadhyay, S., Chaudhuri, R. K., 2008. Molecular applications of state-specific multireference perturbation theory to HF, $H_2O, H_2S, C_2, and N_2$ molecules. *J. Chem. Phys.* 129, 024108.

[17] Sinha Mahapatra, U., Chattopadhyay, S., Chaudhuri, R. K., 2009. Application of state-specific multireference Møller-Plesset perturbation theory to nonsinglet states. *J. Chem. Phys.*, 130, 014101.

[18] Hubač, I., Mach, P., Wilson, S., 2002. On the application of Brillouin-Wigner perturbation theory to multireference configuration mixing. *Mol. Phys.* 100, 859-863.

[19] Mao, S., Cheng, L., Liu, W., Mukherjee, D., 2012. A spin-adapted size extensive state-specific multi-reference perturbation theory. I. Formal developments. *J. Chem. Phys.* 136, 024105.

[20] Chattopadhyay, S., Chaudhuri, R. K., Sinha Mahapatra, U., 2015. State specific multireference perturbation theory with improved virtual orbitals: Taming the ground state of $F_2, Be_2, and N_2$. *J. Comput. Chem.* 36, 907-925.

[21] Sinha Ray, S., Ghosh, A., Chattopadhyay, S., Chaudhuri, R. K., 2016. Taming the Electronic Structure of Diradicals through the Window of Computationally Cost Effective Multireference Perturbation Theory. *J. Phys. Chem. A* 120, 5897-5916.

[22] Sinha Ray, S., Ghosh, P., Chaudhuri, R. K., Chattopadhyay, S., 2017. Improved virtual orbitals in state specific multireference perturbation theory for prototypes of quasidegenerate electronic structure. *J. Chem. Phys.* 146, 064111.

[23] Kurashige, Y., Yanai, T., 2013. Second-order perturbation theory with a density matrix renormalization group self-consistent field reference function: Theory and application to the study of chromium dimer. *J. Chem. Phys.* 135, 094104; Sharma, S., Chan, G. K.-L., 2013. Communication: A flexible multi reference perturbation theory by minimizing the Hylleraas functional with matrix product states. *J. Chem. Phys.* 141, 111101.

[24] Li, C., Evangelista, F. A., 2015. Multireference driven similarity renormalization group: A second-order perturbative analysis. *J. Chem. Theory Comput.* 11, 2097-2108.

[25] Nakano, H., 1993. Quasidegenerate perturbation theory with multi configurational self-consistent-field reference functions. *J. Chem. Phys.* 99, 7983-7992; Granovsky, A. A., 2011. Extended multi-configuration quasi-degenerate perturbation theory: The new approach to multi-state multi-reference perturbation theory. *J. Chem. Phys.* 134, 214113.

[26] Finley, J., Malmqvist, P.-Å., Roos, B.O., Serrano-Andrés, L., 1998. The multi-state CASPT2 method. *Chem. Phys. Lett.* 288, 299-306; Song, C., Martínez, T. J., 2020. Reduced scaling extended multi-state CASPT2 (XMS-CASPT2) using supporting subspaces and tensor hypercontraction. *J. Chem. Phys.* 152, 234113 (1-20).

[27] Angeli, C., Borini, S., Cestari, M., Cimiraglia, R., 2004. A quasidegenerate formulation of the second order n-electron valence state perturbation theory approach. *J. Chem. Phys.* 121, 4043-4049.

[28] Roos, B. O., Andersson, K., 1995. Multiconfigurational perturbation theory with level shift - the Cr_2 potential revisited. *Chem. Phys. Lett.* 245, 215-223.

[29] Witek, H. A., Choe, Y.-K., Finley, J. P., Hirao, K., 2002. Intruder state avoidance multireference MllerPlesset perturbation theory. *J. Comput. Chem.* 23, 957-965.

[30] Potts, D. M., Taylor, C. M., Chaudhuri, R. K., Freed, K. F., 2000. The improved virtual orbital-complete active space configuration interaction method, a packageable efficient ab initio many-body method for describing electronically excited states. *J. Chem. Phys.* 114, 2592-2600.

[31] Chaudhuri, R. K., Freed, K. F., 2005. Relativistic effective valence shell Hamiltonian method: Excitation and ionization energies of heavy metal atoms. *J. Chem. Phys.* 122, 204111.

[32] Gagliardi, L., Truhlar, D. G., Li Manni, G., Carlson, R. K., Hoyer, C. E. Bao, J. L., 2017. Multiconfiguration Pair-Density Functional Theory: A New Way to Treat Strongly Correlated Systems. *Acc. Chem. Res.* 50, 66-72 and references therein.

[33] Li Manni, G., Carlson, R. K., Luo, S., Ma, D., Olsen, J., Truhlar, D. G., Gagliardi, L., 2014. Multiconfiguration pair-density functional theory. *J. Chem. Theory Comput.* 10, 3669-3680.

[34] Manna, S., Sinha Ray, S., Chattopadhyay, S., Chaudhuri, R. K., 2019. A simplified account of the correlation effects to bond breaking processes: The Brillouin-Wigner perturbation theory using a multireference formulation. *J. Chem. Phys.* 151, 064114; Chattopadhyay, S. 2020. Investigation of multiple bond dissociation using Brillouin-Wigner perturbation with improved virtaul orbitals. *J. Phys. Chem. A* 124, 1444-1463.

[35] Jeziorski, B., Monkhorst, H. J., 1981. Coupled-cluster method for multideterminantal reference states. *Phys. Rev. A: At., Mol., Opt. Phys.* 24, 1668-1689.

[36] Pahari, D., Chattopadhyay, S. K., Das, S., Mukherjee, D., U. Sinha Mahapatra, In: Dykstra, C. E., Frenking, G., K. S. Kim, K. S., Scuseria, G. E., Editors. *Theory and Applications of Computational Chemistry: The First 40 Years in Quantum Chemistry.* Amsterdam, Elsevier, 2005; 581.

[37] See www.emsl.pnl.gov/forms/basisform.html. Schuchardt, K. L.;Didier, B. T.; Elsethagen, T.; Sun, L.; Gurumoorthi, V.; Chase, J.; Li, J.; Windus, T. L. Basis Set Exchange: A Community Database for Computational Sciences. *J. Chem. Inf. Model.* **2007**, *47*, 104-1052 and references therein.

[38] Sand, A.M., Schwerdtfeger, C.A., Mazziotti, D.A., 2012. Strongly correlated barriers to rotation from parametric two-electron reduced-density-matrix methods in application to the isomerization of diazene. *J. Chem. Phys.* 136, 034112.

[39] Angeli, C., Cimiraglia, R. and Hofmann, H.J., 1996. On the competition between the inversion and rotation mechanisms in the cis-trans thermal isomerization of diazene. *Chem. Phys. Lett.* 259, 276-282.

[40] Mach, P. Mášik, J., Urban, J., Hubač, I., 1998. Single-root multireference Brillouin-Wigner coupled-cluster theory. Rotational barrier of the N_2H_2 molecule. *Mol. Phys.* 94, 173-179.

[41] Biczysko, M., Poveda, L.A., Varandas, A.J.C., 2006. Accurate MRCI study of ground-state N_2H_2 potential energy surface. *Chem. Phys. Lett.* 424, 46-53.

[42] Musiał, M., Lupa, Ł., Szopa, K., Kucharski, S.A., 2012. Potential energy curves via double ionization potential calculations: example of 1, 2-diazene molecule. *Struct. Chem.* 23, 1377-1382.

[43] Chaudhuri, R.K., Freed, K.F., Chattopadhyay, S., Sinha Mahapatra, U., 2008. Potential energy curve for isomerization of N_2H_2 and C_2H_4 using the improved virtual orbital multireference MllerPlesset perturbation theory. *J. Chem. Phys.* 128, 144304.

[44] Mahapatra, U.S., Chattopadhyay, S., 2011. Evaluation of the performance of single root multireference coupled cluster method for ground and excited states, and its application to geometry optimization. *J. Chem. Phys.* 134, 044113.

[45] Mahapatra, U.S., Chattopadhyay, S., Chaudhuri, R.K., 2011. Secondorder statespecific multireference Møller Plesset perturbation theory: Application to energy surfaces of diimide, ethylene, butadiene, and cyclobutadiene. *J. Comp. Chem.* 32, 325-337.

[46] Demaison, J., Hegelund, F., Bürger, H., 1997. Experimental and *ab initio* equilibrium structure of trans-diazene HNNH. *J. Mol. Struct.* 413, 447; Hegelund, F., Bürger, H., Polanz, O., 1994. The High-Resolution Infrared Spectrum of the v4, v5, and v6 Bands of trans-Di-imide Revisited. *J. Mol. Spectrosc.* 167, 1.; Sylwester, A.P., Dervan, P.B., 1984. Low-temperature matrix isolation of the 1, 1-diazene H_2NN. Electronic and infrared characterization. *J. Am. Chem. Soc.* 106, 4648.; Biehl, H., Stuhl, F., 1994. Vacuumultraviolet photolysis of N2H2: Generation of NH fragments. *J. Chem. Phys.* 100, 141.

[47] Song, L., Liu, M., Wu, W., Zhang, Q., Mo, Y., 2005. Origins of rotational barriers in hydrogen peroxide and hydrazine. *J. Chem. Theory. Comput.* 1, 394-402.

[48] Dyczmons, V., 2000. Six structures of the hydrazine dimer. *J. Phys. Chem. A* 104, 8263-8269.

[49] Liu, S., Govind, N., Pedersen, L.G., 2008. Exploring the origin of the internal rotational barrier for molecules with one rotatable dihedral angle. *J. Chem. Phys.* 129, 094104.

[50] Łodyga, W., Kreglewski, M., Makarewicz, J., 1997. The InversionTorsion Potential Function for Hydrazine. *J. Mol. Spectrosc.* 183, 374-387.

[51] Łodyga, W., Makarewicz, J., 2012. Torsion-wagging tunneling and vibrational states in hydrazine determined from its ab initio potential energy surface. *J. Chem. Phys.* 136, 174301.

[52] Kasuya, T., Kojima, T., 1963. Internal motions of hydrazine. *J. Phys. Soc. Jpn.* 18, 364-368.

[53] Tsunekawa, S., 1976. Microwave Spectrum of Hydrazine-1,2-d_2. *J. Phys. Soc. Jpn.* 41, 2077-2083.

[54] Kohata, K., Fukuyama, T., Kuchitsu, K., 1982. Molecular structure of hydrazine as studied by gas electron diffraction. *J. Phys. Chem.* 86, 602-606.

[55] Giguere, P.A., Liu, I.D., 1952. On the infrared spectrum of hydrazine. *J. Chem. Phys.* 20, 136-140.

[56] Durig, J.R., Griffin, M.G., MacNamee, R.W., 1975. Raman spectra of gases. XV: Hydrazine and hydrazined$_4$. *J. Raman Spectrosc.* 3, 133-141.

[57] Sinha Ray, S., Sinha Mahapatra, U., Chaudhuri, R. K., Chattopadhyay., S., 2017. Combined complete active space configuration interaction and perturbation theory applied to conformational energy prototypes: Rotation and inversion barriers. *Comp. Theor. Chem.* 1120, 5678.

INDEX

A

acid, 5
additive, viii, 2, 4, 27, 28, 32, 46
and phase, vii, viii, 1, 3, 13, 14, 17, 18, 28, 29, 30, 31, 34, 43, 46, 48
applications of electronic structure theory, 131, 145
atomic orbitals, ix, 99, 100, 101, 102, 103, 109, 110, 112, 115, 120, 125, 141
atomic-orbital (AO) basis functions, 134, 136
atoms-in-molecules (AIM), 2, 13, 45, 52

B

base, 1, 5
basics of electronic structure theory, 131, 151
bridge interactions, 2, 54

C

charge-transfer (CT), vii, viii, 1, 2, 5, 10, 24, 25, 32, 34, 35
chemical potentials, 6, 9, 10, 11, 12, 34, 46
chemical potentials equalized, 12
chemical reactivity theory, 2, 45, 57
chemical-potential, 46
classical entropy, 19
classical it measures, 3, 4, 17
closed, 4, 6, 9, 11, 29, 34, 35, 39, 40, 41, 42, 45, 46, 48, 50, 152, 165, 168
complex (vector) entropy, 20
complex entropy, 33, 55
components, x, 13, 16, 17, 18, 20, 30, 37, 38, 39, 40, 41, 43, 54, 55, 155, 164
composite system, 11, 23, 34, 40, 41, 137
conditional probabilities, 37, 38, 39
continuity equation(s), 13, 14, 18
continuity relations, 2, 3, 16, 45
contragradience (CG), 2
coupled cluster (CC) methods, 135, 136, 137, 138, 158, 168, 169, 170, 174
CT phenomena, 32
current densities, 13

D

wave function theory, 131

density functional theory (DFT), ix, 2, 23, 30, 43, 54, 55, 56, 57, 131, 135, 136, 141, 144, 146, 147, 148, 150, 151, 178
density operator, 23, 24, 25, 40, 46
density-per-particle, 43
descriptors, vii, viii, 1, 3, 5, 6, 21, 25, 27, 28, 29, 32, 33, 41, 42, 46, 52, 53, 55, 56, 57
determinant, 30, 36, 39, 46, 134, 138, 153, 172
disentangled, 3, 4, 6, 29, 42, 45
distinguishable, 30
donor-acceptor, vii, viii, 1, 3, 6, 8
donor-acceptor systems, v, vii, viii, 1, 2, 3, 4, 8
dynamics, 13, 14, 52, 56, 132, 139, 142, 149, 156

E

effective velocity, 3, 14, 15, 16
electron communications, 12, 53
electron localization function (ELF), 2, 53, 54
electron reservoir(s), 7, 8, 11, 23, 24, 34, 40, 44, 46
electronic communications, 2, 3, 45
electronic current, 19, 37
electronic structure, vii, viii, ix, x, 4, 7, 16, 24, 45, 51, 52, 55, 56, 59, 60, 62, 64, 67, 78, 85, 90, 95, 96, 99, 100, 101, 102, 104, 108, 109, 110, 114, 119, 120, 121, 122, 125, 127, 131, 132, 133, 135, 137, 139, 142, 143, 144, 145, 146, 150, 151, 152, 153, 156, 160, 166, 173, 177
electronic wavefunctions, 22, 45, 134, 146
electrons, 3, 4, 5, 8, 10, 11, 17, 22, 23, 24, 25, 26, 27, 30, 31, 35, 39, 41, 43, 44, 45, 68, 72, 73, 78, 85, 89, 91, 102, 104, 107, 110, 111, 112, 114, 115, 116, 118, 119, 120, 121, 125, 132, 133, 134, 136, 138, 139, 141, 146, 161, 170
energetic principle, 46
ensemble description, 11, 24, 29
entangled, 3, 4, 6, 30, 32, 33, 40, 42, 45, 46, 56
entropy, 3, 16, 17, 19, 20, 27, 28, 29, 31, 32, 33, 42, 44, 45, 46, 47, 52, 53, 55, 56
entropy of shannon, 19
equalization, 8, 10, 33, 46
equalization of chemical potentials, 8
equalization of phases, 33
equalized, 6, 9, 11, 35, 40, 41
equidensity orbital(s) (EO), 4, 30, 31, 32, 34, 35, 36, 37, 38, 39, 40, 41, 42, 46, 47, 51, 58
equilibrium, 3, 7, 9, 10, 11, 12, 13, 22, 24, 25, 31, 34, 35, 36, 40, 41, 42, 43, 44, 45, 46, 47, 48, 50, 51, 55, 56, 65, 75, 81, 82, 84, 121, 142, 180
equilibrium EO, 36
equilibrium phase, 36, 42, 51
equilibrium probabilities, 24, 41
equilibrium stage, 10
external-closeness, 9
externally-closed, viii, 2, 22, 24
extremum principles, 3
extremum rules, 45

F

Fisher, viii, 2, 3, 4, 16, 20, 51, 53, 54
Fisher information, 2, 51, 53, 54
flux, 19

G

geometric entropy, 28, 33
geometric measure, 3, 21
geometric measure of the resultant entropy, 21

Index

global, viii, 2, 3, 4, 6, 9, 12, 17, 19, 21, 33, 35, 46
global-entropy, 21
gradient information/entropy, 3, 16, 17, 18, 20, 21, 22, 33, 41, 46
gradient measures, 20
grand-ensemble, viii, 2, 3, 4, 5, 25, 27, 40, 44, 46

H

Harriman-Zumbach-Maschke (HZM) construction, 30, 31, 36, 37, 38, 39, 46
Hartree-Fock (HF) methods, 134, 138, 141, 161, 165, 176
Hilbert space, x, 155, 168
HZM construction, 31, 38, 46

I

indistinguishable electrons, 30, 39
information, v, vii, viii, 1, 2, 3, 16, 17, 18, 19, 20, 21, 22, 27, 29, 31, 32, 34, 42, 45, 46, 47, 51, 52, 53, 55, 56, 57, 58, 62, 86, 127, 144, 167
information criteria, 45
information production, 3, 18
information source, 19
information theory (IT), vii, viii, 1, 2, 4, 17, 20, 27, 28, 41, 45, 51, 52, 55
information-equilibrium, 46
inter-reactant CT, 11
intruder states, 159, 164, 174
isomerization process, 168, 169

J

Jacobian, 37, 38
Jacobian determinant, 38

K

kinetic-energy, 3, 16, 17, 19, 22, 41, 42, 44, 45, 56, 119

L

layered structure, 60, 67, 83
linear-combination-of-atomic-orbital (LCAO), 102, 109, 112, 120, 134
logarithmic derivative, 22, 49

M

magnetic dilution, vii, viii, 59, 60, 62, 65, 67, 70, 72, 77, 78, 79, 81, 86, 90
magnetic susceptibility, vii, viii, 59, 60, 62, 63, 65, 66, 68, 69, 70, 71, 72, 73, 74, 80, 81, 82, 84, 88, 89, 90, 95
many-body perturbation theory (MBPT), 135, 136, 137, 138
maximum-entropy principles, 3
minimum overall information, 36
minimum-energy, 3, 55
minimum-energy principle, 3
mixed quantum state(s), 23, 25
mixed state, 24, 40, 44
Mjøller-Plesset (MP) methods, 138
modulus, viii, 1, 3, 13, 14, 15, 17, 18, 22, 23, 30, 31, 32, 34, 42, 43, 45, 46, 122, 123, 124
modulus and phase components, 13, 23
modulus component, 15
molecular spin-orbitals (MO), 102, 109, 112, 115, 118, 120, 134
mutual-closeness, 8
mutually- and externally-closed, 23
mutually-open, viii, 2, 4, 6, 10, 11, 12, 22, 24, 30, 31, 32, 33, 34, 35, 40, 46, 48

Index

N

nanomaterials, v, vii, viii, 93, 99, 100
nonadditive, 2, 42, 54

O

of, 24, 45
opaque division wall, 4, 42, 48
open, viii, x, 2, 3, 4, 5, 7, 8, 12, 22, 23, 24, 25, 30, 32, 34, 35, 40, 41, 42, 44, 45, 50, 156
open reactants, 5, 8
open systems, viii, 2, 22
orbital communication theory (OCT), 2, 53
orthogonality phase, 38
overall entropy, 20, 21, 27, 57

P

Parameterized Model 3, 100
perovskite, viii, 59, 60, 63, 67, 74, 77, 78, 79, 81, 82, 85, 91, 92
phase, viii, 1, 2, 3, 4, 14, 15, 16, 17, 18, 19, 22, 25, 29, 30, 31, 33, 34, 35, 36, 40, 41, 42, 43, 45, 46, 47, 50, 55, 65, 128, 129, 132, 133, 135, 137, 140, 142, 147, 151, 153
phase components, 45
phase equalization, 4, 29, 36
phase Laplacian, 15, 47, 48
phase-continuity, 15, 18
phase-dynamics, 15
phase-equalization, viii, 2, 33, 34, 35, 46
phase-flow, 15
phase-flux, 15, 16
phase-production, 15, 18, 42
phases, 3, 34, 35, 40, 46, 47, 67, 101, 126, 140
phase-source, viii, 1, 42

polarized, 7, 9, 10, 11, 12, 23, 29, 30, 31, 34, 35, 45
polarized reactive system, 9, 45
polarized system, 30, 35
probabilities, 24, 29, 40, 44, 46, 52, 142
probability, vii, viii, 1, 2, 3, 9, 10, 12, 13, 14, 15, 16, 17, 19, 21, 22, 24, 25, 27, 28, 30, 31, 32, 34, 35, 36, 37, 38, 40, 42, 43, 45, 46, 47, 50, 51, 133
probability, 14
probability-current, 15, 16
production, viii, 1, 16, 19, 158
promolecular reference, 35, 45
promolecular stage, 9
promolecular state, 9
promolecular system, 12

Q

quantum monte carlo (QMC) methods, 138, 141
quasidegeneracy, vii, x, 155

R

reactants, viii, 2, 4, 8, 10, 12, 22, 29, 30, 31, 33, 34, 35, 45
reaction stages, vii, viii, 1, 4, 7, 13, 45
reactive system, viii, 2, 3, 4, 5, 6, 9, 10, 11, 12, 13, 22, 23, 24, 27, 28, 29, 32, 34, 36, 46, 55
reactivity criteria, 56
reactivity theory, 4
relativistic effective core potentials (RECPs), 134
reservoir, 5, 9, 12, 24, 46
resultant, v, vii, viii, 1, 2, 3, 5, 9, 13, 16, 17, 18, 19, 20, 21, 24, 27, 28, 31, 32, 34, 36, 41, 42, 44, 45, 46, 47, 55, 56, 58
resultant descriptors, 17

resultant entropy/information, 2, 3, 21, 27, 32, 45
resultant gradient entropy, 20
resultant gradient-information, viii, 1, 3, 5, 17, 18, 19, 20, 34, 44, 45, 47, 56
resultant information/entropy, vii, viii, 1, 28, 41, 42, 55, 58
resultant IT descriptors, 3
resultant-entropy functionals, 32
rotational barrier, 166, 169, 170, 172, 173, 174, 175, 179, 180

S

scalar entropy, 19, 32
Schrödinger equation (SE), 3, 13, 14, 43, 45, 48, 49, 50, 132, 133, 135, 137, 139
Shannon, viii, 1, 2, 3, 4, 17, 20, 21, 51, 92
Shannon entropy, 21
similarity transformation, 26
slater function, ix, 100, 101, 125
sources, 96
spinels, 60, 85, 86, 87, 88, 89, 93
stages of chemical processes, 6
stages of chemical reactions, 3
state specific, v, 155, 160, 169, 177
stockholder division, 2
structure of, 45
subsystems, viii, 2, 3, 4, 5, 6, 7, 8, 9, 10, 11, 12, 13, 23, 24, 27, 28, 29, 30, 31, 32, 33, 34, 35, 40, 42, 45, 46, 48

system, 3, 4, 5, 6, 7, 12, 20, 22, 23, 24, 30, 32, 33, 34, 37, 39, 40, 41, 42, 43, 44, 45, 46, 48, 50, 51, 65, 67, 68, 70, 72, 79, 81, 84, 85, 90, 102, 103, 104, 111, 112, 115, 116, 120, 122, 125, 134, 135, 139, 141, 165, 166, 167, 170, 171

T

thermodynamic conditions, 24, 40, 44
thermodynamic principles, 45
torsional surface, 168, 171, 174

V

vanishing divergence, 15
variational principle(s), 3, 37, 45
vertical/horizontal, 8, 9, 34, 35
virial theorem, 3, 45, 55

W

wavefunction phase, 3
wavefunction(s), 3, 13, 15, 22, 29, 30, 31, 42, 43, 45, 46, 47, 48, 133, 134, 135, 136, 137, 138, 139, 141, 142, 149, 166, 173